The Pesticide Problem

The widespread use of chemicals to control pests has resulted in adverse effects for both wildlife and humans. Originally published in 1967, this title seeks to clearly explain the key issues for understanding public policy in the pesticide problem. Authors Headley and Lewis provide simple clarification of the economic issues involved in creating public policy for pest control and present how policy formation for pesticides will be improved by further economic analysis. This title is a valuable and relevant resource for students interested in environmental studies, especially the impact of public policy making on the environment.

W0235012

The Pesticide Problem

An Economic Approach to Public Policy

J.C. Headley and J.N. Lewis

First published in 1967
by Resources for the Future, Inc.

This edition first published in 2016 by Routledge
2 Park Square, Milton Park, Abingdon, Oxon, OX14 4RN
and by Routledge
711 Third Avenue, New York, NY 10017

Routledge is an imprint of the Taylor & Francis Group, an informa business

© 1967, Resources for the Future, Inc.

Publisher's Note
The publisher has gone to great lengths to ensure the quality of this reprint but points out that some imperfections in the original copies may be apparent.

Disclaimer
The publisher has made every effort to trace copyright holders and welcomes correspondence from those they have been unable to contact.

A Library of Congress record exists under LC control number: 66028503

ISBN 13: 978-1-138-95899-9 (hbk)
ISBN 13: 978-1-315-66088-2 (ebk)
ISBN 13: 978-1-138-95901-9 (pbk)

THE PESTICIDE PROBLEM:

AN ECONOMIC APPROACH
TO PUBLIC POLICY

by
J. C. Headley
and
J. N. Lewis

RESOURCES FOR THE FUTURE, INC.
1755 Massachusetts Avenue, N.W., Washington, D.C. 20036

Distributed by
THE JOHNS HOPKINS PRESS
Baltimore and London

Resources for the Future is a nonprofit corporation for research and education in the development, conservation, and use of natural resources and the improvement of the quality of the environment. It was established in 1952 with the co-operation of the Ford Foundation. Part of the work of Resources for the Future is carried out by its resident staff; part is supported by grants to universities and other nonprofit organizations. Unless otherwise stated, interpretations and conclusions in RFF publications are those of the authors; the organization takes responsibility for the selection of significant subjects for study, the competence of the researchers, and their freedom of inquiry.

This book is one of RFF's studies in quality of the environment, which are directed by Allen V. Kneese. J. C. Headley is associate professor of agricultural economics at the University of Missouri; J. N. Lewis is professor of agricultural economics at the University of New England, Australia, and was a visiting professor at the University of Illinois while working on the study. The manuscript was edited by Elizabeth S. Reed.

RFF staff editors: Henry Jarrett, Vera W. Dodds, Nora E. Roots, Tadd Fisher.

Preface

The impact of resource use upon the quality of resource commodities and the physical environment has become of major concern throughout the world. That pollution of the air, land, and water, together with degradation of the urban and rural landscape, have assumed major proportions in the United States is reflected in both the action taken and the research programs begun by public and private agencies. These environmental quality problems pose unusually challenging and complex policy issues, in part because their consequences are so difficult to measure and in part because our economic and political institutions have not been geared to deal with them.

Several years ago Resources for the Future decided to give special emphasis to the study of environmental quality problems. As in other of its fields of study, the underlying objective was to develop analyses and information which would provide an improved foundation for public and private policy formation. Since then, a substantial amount of research has been initiated and a forum has been held at which experts who have been studying environmental quality problems discussed the nature of their research. The publications that have resulted include: *Environmental Quality in a Growing Economy* (papers presented at the 1966 RFF Forum, edited by Henry Jarrett); *Quality of the Environment: An Economic Approach to Some Problems in Using Land, Water, and Air,* by Orris C. Herfindahl and Allen V. Kneese; *The Economics of Regional Water Quality Management,* and *Water Pollution: Economic Aspects and Research Needs,* both by Allen V. Kneese. The present study is the first in the RFF series to deal specifically with the use of pesticides.

The study is based on the premise that a rational approach to

v

policy development with regard to the use of pesticides must be concerned with the values—the costs and the returns—associated with such use. No matter what scientific advances are likely to be made, the use of pesticides will involve benefits, costs, and probably some uncertainty and risk. If we are to develop wise policies, the benefits, the measurable costs, and the costs of avoiding uncertainty and risk associated with alternative policies must be weighed and considered objectively and carefully; no other course can be considered reasonable or scientific. The economics discipline appears to offer the most appropriate methodology for weighing these values in a rigorous and systematic way.

The study provides a foundation for future policy research on the use of pesticides. It offers a broad picture of the nature and extent of present use and the policy issues we face. It also provides a conceptual framework for policy analyses, a preliminary assessment of the consequences of pesticides, and suggested approaches to further research on this subject. The study is not and could not be a definitive analysis of the many factors that bear upon the formulation of future policy. It does, however, take an essential first step in a program of research that must be undertaken by public and private institutions to provide the understandings necessary for progressive development of reasonable policies governing the use of pesticides.

Irving K. Fox, Vice President
Resources for the Future, Inc.

Acknowledgments

A study such as this engages a number of issues in numerous disciplines. Consequently, we are indebted to many people for their assistance in the course of our inquiry.

The encouragement and guidance given us by Irving K. Fox and Allen V. Kneese, of Resources for the Future, Inc., both in planning the study and while it was in preparation, was invaluable. Their constructive suggestions on early drafts of the manuscript contributed in many ways to its completion.

Many people outside RFF provided information, reviewed the manuscript, and made useful suggestions. Particular acknowledgment goes to Velmar W. Davis, Farm Production Economics Division, Economic Research Service, U.S. Department of Agriculture; Lyle P. Fettig, Department of Agricultural Economics, University of Illinois; John L. George, Department of Forestry and Wildlife, Pennsylvania State University; S. A. Hall and E. F. Knipling, Entomology Research Division, Agricultural Research Service, U.S. Department of Agriculture; Robert S. Roe, Bureau of Standards and Evaluation, Food and Drug Administration, U.S. Department of Health, Education, and Welfare; Robert L. Rudd and Kenneth E. F. Watt, Department of Zoology, University of California (Davis); and Burton A. Weisbrod, Department of Economics, University of Wisconsin. The constructive criticisms and encouragement offered by these people were most helpful.

Even though we solicited and received assistance from the individuals mentioned, the responsibility for accuracy in concept, fact, and presentation remains with the authors.

J. C. HEADLEY
J. N. LEWIS

Contents

Tables

Tables

Introduction

In attempts to manage his environment so as to provide himself with the necessary elements of life and pleasant surroundings, man has found himself in competition with various other forms of life. Notable among these are certain insects and plants that either destroy what man has produced, compete for resources needed by man in production of food and fiber, or contaminate the environment in such a way that man finds it unhealthy or unattractive.

These competitive insects and plants have come to be known as "economic pests." Because of the competitive nature of these pests, man has seen fit to allocate resources toward their reduction and, sometimes, extermination.

Attempts to control such economic pests as certain insects and certain plants called weeds have taken at least four different forms. First, mechanical methods such as cultivation are used to control weeds. Similarly, barriers to movement, destruction of the habitat, and trapping have been used to control insects. Second, cultural methods such as the timing of planting, deep plowing, and the rotation of crops have been employed. Third, biological methods making use of diseases known to affect the pests, development of resistant crop varieties, and the use of predator or parasitic insects have been found effective. Fourth, a variety of chemicals have been developed which kill the undesirable plant or insect and thereby reduce the competitive species.

The use of chemicals to control economic pests has grown in recent years both in the amount and the diversity of materials used. In fact, the increase in the scale of use of chemicals to control pests, the reports of such adverse effects as fish and wildlife kills resulting from their use, and the discovery of DDT in the adipose tissue of humans have caused thoughtful people to search for the facts concerning large-scale usage of pesticides and the resulting side effects. Considerable attention has been focused on the development of policies that will combine technology and institutions to achieve human goals and values with some degree of optimality.

The search for information and attempts to develop policies have led to a controversy characterized by heated discussion, the appearance of such books as the one by the late Rachel Carson [1], and the appropriation of considerable funds by Congress for research concerning pesticides. Unfortunately, the controversy has been marked by a certain amount of defensiveness on the part of conservationists, agriculturalists, scientists, and the chemical industry. Consequently, the layman has been exposed to many statements through the various media—a good percentage of which have been launched from a defended position. The difficulty in obtaining a calm and dispassionate account of the situation has made the job of policy making more difficult.

Our objective in the discussion that follows is to develop a clearer understanding of the issues for public policy in the pesticide problem, to suggest an approach to policy formation based on economic analysis, to review what is known about the technical relationships between pesticides and environmental quality, and to propose research approaches that will provide guidance for future pest control policy.

Because of a tendency to oversimplify, the need for clarification of the economic issues in public policy for pest control is great. More often than not, the approach to the question and the solutions suggested have tended to maximize certain values and disregard others. That many of the issues involved are economic in nature or have important economic implications is only imperfectly realized by many scientific writers on the subject. One leading pest control specialist has suggested that the necessity in pest control to make economic choices between conflicting objectives is unacceptable. He states that "if certain pest control measures are found to cause serious adverse effects, it then becomes necessary to weigh gains against potential risks. This is not a desirable situation. Such problems can lead only to disagreement and controversy. Therefore, every possible effort is justified in attempts to meet pest problems without the necessity of accepting any losses to other values" [2]. This statement represents a hope that, if sufficient investment is made in research, choices will not have to be made and economic problems will disappear.

In contrast with the above position, it is suggested that the obvious dependence of modern agriculture upon pesticides is sufficient reason for doing nothing to restrict their use, since it is clear that the gains in agricultural productivity outweigh losses to other values. This does not seem to be an acceptable solution to the problem, nor is there any answer in asserting that other environmental changes wrought by man, such as urbanization, land clearing, drainage of wet lands, and highway construction, have done far more to affect wildlife than pesticides, or that motor vehicles have killed more people than parathion.

Proceeding on the premise that the questions raised by the addition of large quantities of pesticides to the environment cannot be satisfactorily resolved by resorting to panaceas or by treating the problem as though it were independent of other elements of technology, we have, in this exploratory study, attempted to place the pesticide problem in perspective and to treat it as part of the more general problem of pollution and environmental quality [3]. Chapter 1 presents the background and nature of the problem, considers the types and amounts of materials used and acreages treated and discusses the economics of spillover hazards resulting from use of pesticides.

While a complete appraisal of the social costs and benefits of pesticides has not been made and perhaps will not soon be made, because of problems with extra market values and deficiencies in the data concerning technical relationships of pesticides to agricultural production, health, and natural resources, it is meaningful to develop a conceptual decision framework for the formulation of policy. A decision framework is presented in Chapter 2. This framework defines an optimal pest control policy as being determined by the maximization of social values. It translates adverse effects from pesticides into economic variables and shows the policy implications of the different values these variables might take.

In Chapter 3, the assessment of the effects of pesticides on agriculture and consumers is examined. Using supply and demand for agricultural products as measures of resource costs and consumer satisfaction respectively, the effects of pesticide technology on resources devoted to agricultural production and on consumer satisfaction are analyzed. Various market conditions are assumed, including the free market and the market under alternative forms of price support and output control. The discussion also includes the conceptual and measurement problems attendant to the assessment. The weaknesses of the tacit models employed in some of the more usual approaches to estimation of the benefits from pesticide use are also revealed.

A review of the potential consequences of pesticide use on agricultural productivity, human health, and fish and wildlife is presented in Chapters 4, 5, and 7. Here we describe as objectively as possible, from available information, the form and incidence of the effects of pesticides in qualitative and, where possible, quantitative terms. Chapter 6 discusses the conceptual and empirical problems related to the evaluation of the effect of pesticides on health, while similar problems with respect to fish and wildlife consequences are treated in Chapter 7.

Finally, in Chapter 8, suggestions for future research are made along with an examination of current institutions providing for regulation of pesticide usage and protection of consumers. Embodied in this section is the expressed need for research co-operation among physical, natural, and social scientists to achieve results leading to policies that consider the broad range of values involved.

The economic analysis of the effects of pesticides is a largely undeveloped area and the disorganized state of the data prevents the formulation and solution of a neat statistical model for fabricating decisions for society. We believe, however, that policy formulation for pesticides will almost certainly be improved by an explicit and systematic consideration of the issues.

Entirely satisfactory public programs are difficult, if not impossible, to develop. It is very unlikely that programs emerging from slogans will ever stand even the weakest test of social optimality. It is equally unlikely that the best policies will result from the delegation of full responsibility to a scientific elite [4], especially one that scorns economic decisions as somehow non-scientific. It is our purpose to make some beginning to the economic appraisal of pesticide usage in the belief that the so-called pesticide problem is largely and inescapably economic in nature.

REFERENCES

1. Rachel Carson, *Silent Spring* (Boston: Houghton Mifflin Co., 1962).
2. E. F. Knipling, *Statement Before the Subcommittee on Reorganization and International Organizations of the Senate Committee on Government Operations,* October 7, 1963, p. 25. Mimeo.
3. Orris C. Herfindahl and Allen V. Kneese, *Quality of the Environment* (Washington, D.C.: Resources for the Future, Inc., 1965), p. 35.
4. American Association for the Advancement of Science, Committee on Science in the Promotion of Human Welfare, "The Integrity of Science," *American Scientist,* Vol. 53, No. 2, June 1965, p. 191.

THE PESTICIDE PROBLEM:

AN ECONOMIC APPROACH TO PUBLIC POLICY

1

Background and Nature of the Pesticide Problem

The hazards associated with pesticides have grown both in magnitude and complexity during the last two decades. Increasing hazards, Knipling suggests [1], arise from the development of many new kinds of chemicals for use as insecticides, herbicides, fungicides, and rodenticides. Many of these new chemicals are very many times more toxic than pesticides previously in use and they can, moreover, be ingested by inhalation and absorption through the skin in addition to the more familiar oral route. Also, the increasing quantities of pesticides employed and the widening range of uses and products to which chemicals are applied only aid the proliferation of the new chemicals and extend the probable hazards.

Prior to the year 1800, compounds of lye, lime, soap, turpentine, tobacco, pyrethrum powder, mineral oil, and arsenic were reportedly used as insecticides [2]. However, the use of Paris green to control the Colorado beetle in 1867 marks the beginning of commercial pesticides. The success of Paris green led to trials with other arsenic compounds and the later commercial use of lead arsenate and calcium arsenate. Until DDT made its appearance in 1945, the use of agricultural insecticides was confined largely to fruits, vegetables, cotton, and a few other high value crops.

With the development of the new synthetic organic pesticides, including the chlorinated hydrocarbons, the organic phosphates, and the carbamates, spraying and dusting operations have spread to field crops, pastures, and forests. Insecticides are widely used in food storage, in the control of insect vectors of disease, in the control of nuisance insects, and in homes and gardens. With the development of herbicides, such as 2,4-D and 2,4,5-T, the use increased rapidly and has become much more diversified. Apart from their use in controlling weeds found in intertilled crops, pastures, and lawns, they have been used extensively in controlling vegetation on roadsides, for land clearing (using fire after treatment of scrub and light timber with chemicals), and for the control of aquatic and other weeds in irrigation and drainage canals.

DDT is the best known of the extensive class of chemicals known as chlorinated hydrocarbons. Others in the group include chlordane, benzene hexachloride, lindane, dieldrin, aldrin, endrin, toxaphene, methoxychlor, and heptachlor.

In addition to high acute and chronic toxicity levels exhibited by some of these compounds, the chlorinated hydrocarbons as a group possess another characteristic which, for many purposes, is very desirable in insecticides. This is the quality of persistence, useful in combating many insect pests. Although some insects, such as aphids, overwinter in the egg state and are extremely vulnerable to such simple ovicides as tar distillates, others are not all exposed to a poison at one and the same time and protection of a crop against them requires treatment with a persistent chemical or repeated use of a nonpersistent chemical. Control of adult mosquitoes calls for the same property. This very quality of persistence, however, has added to the hazards involved, although this is by no means a new phenomenon in pesticides. The arsenic compounds are also persistent to a very high degree.

A second major group of pesticidal chemicals consists of the organic phosphates, including among many others, parathion, malathion, and tetraethyl pyrophosphate (TEPP). According to one author [3], there has been a strong trend to the organophosphates which, "as a class, comprise by far the greatest number of new candidate materials that are being tested and synthesized as insect control agents." The great advantage of the organophosphates lies in their varied properties. Some, such as TEPP, hydrolize and leave no residue after use. Others are systemic in growing plants or animals.

A lesser trend to a new class of chemicals, the carbamates, has been stimulated by the development of insect resistance to the chlorinated hydrocarbons and organophosphates and by the search for control agents which are relatively nontoxic to warm-blooded animals and which do not accumulate as residues in animal tissues. The carbamate insecticide, Sevin, has this desirable property although some of the carbamates, like many of the organophosphates, are very toxic and inhibit the supply of cholinesterase, a substance essential to the functioning of the nervous system in mammals. In addition, some of the carbamates have growth-regulating effects on plants [3]. This property enables their use as selective herbicides, but is undesirable in an agricultural insecticide.

A major technical problem in pest control is the development of resistance to insecticides among insect populations. Resistance to pesticides is not a new problem. It has been greatly intensified, however, by the very effectiveness of the newer type chemicals. Because of the extremely high rates of kill obtained, the process of selection for resistance is enormously accelerated. As a consequence, strains of an insect pest resistant to a particular chemical may emerge within a few years. Any resistant strains that did develop with the older type of pesticides usually took several decades to appear. The problem is accentuated by cross-resistance, that is, by the fact that resistance developed to one chemical sometimes confers or predisposes resistance to other chemicals of the same group. Chemical control of insect pests, therefore, acquires a temporary character involving continuous efforts to develop new agents to replace those which are rendered obsolete by insect resistance. Like Lewis Carroll's Red Queen, the chemical pesticide industry has to do a great deal of running to stay in the same place.

This tendency of chemicals to become obsolete has been held by some to indicate the superiority of biological controls where the job stays done for longer periods of time and is sometimes even permanent. However, there are dangers in evaluating pest control methods on the basis of this criterion alone. Much scientific research is defensively oriented and consists of solving problems which have grown out of achievements based on earlier research findings. Creation of second-round problems is a general phenomenon inseparable from technological progress.

It would also be questionable, from a policy standpoint, to place overriding value on the criterion of nontoxicity to vertebrates, although this is a desired property in certain situations such as household controls of insects. A chemical may be relatively harmless to birds and mammals and still occasion serious side effects if it has, for example, broad spectrum toxicity to insects and so eliminates pests and beneficial insects indiscriminately. Minor damage to wildlife may consequently be preferable to the near elimination of honey bees and other pollinators.

Trends in Output and Domestic Disappearance of Pesticides

Trends in U.S. manufacture and domestic utilization of major pesticides are shown in Tables 1 and 2 respectively. Data are not separately available for many major chemicals, since aggregation is necessary to preserve confidential information on the output of individual firms. As a result, it is impossible to derive totals for main classes of chemical compounds. Nevertheless a number of major trends may be discerned.

Following the introduction of DDT, there was a major decline in the production and domestic disappearance of the arsenicals (Tables 1 and 2). This trend has continued and only a few million pounds of lead and calcium arsenate are currently used. This is in marked contrast with 1939 when the production of these two materials totaled 100 million pounds. (See Table 1.) There has also been a considerable decline in the use of botanically derived materials, principally rotenone and pyrethrum. Data reported by the U.S. Bureau of the Census show that rotenone imports declined from 4.8 million pounds to 3.7 million pounds from 1958 to 1962. Census data also show a declining value of pyrethrum of about $1 million from 1960 to 1962. This decline in value is not due to a decline in prices, which actually rose slightly over the period. The once popular nicotine sulfate has practically disappeared from use.

The introduction of synthetic organic pesticides has, however, led to far more than a replacement of the old by the new materials. Despite the fact that a pound of DDT or malathion goes a lot further than a pound of the older materials, the total quantity of pesticides produced and sold has increased rapidly. Sales of synthetic organic pesticides, as reported by the U.S. Tariff Commission, increased from 279 million pounds in 1954 to about 634 million pounds in 1962 [4]. These figures indicate the rapid expansion in the adoption of the new materials. The

Table 1. Production of Selected Pesticides, United States (1,000 lb.)

Item	1939	1945	1950	1955	1960	1963
Calcium arsenate	41,349	25,644	45,348	3,770	6,590	n.a.
Lead arsenate	59,569	70,522	39,434	14,776	10,062	n.a.
White arsenic	44,686	48,698	26,546	—	—	—
Copper sulfate	134,032	251,000	174,600	156,176	116,000	83,272
Aldrin-toxaphene group[a]	—	—	—	77,025	90,671	105,986
Benzine hexachloride[b]	n.a.	n.a.	76,698	56,051	37,444	6,778
DDT	n.a.	33,243	78,150	129,693	164,180	178,913
Methyl bromide	—	—	—	9,222	12,659	17,394
Methyl parathion	—	—	—	—	11,794	15,999
Parathion	—	—	n.a.	5,168	7,434	c
Nabam	—	—	—	—	2,978	2,420
2,4-D acid	n.a.	917	14,156	34,516	36,185	46,312

[a] Includes the chlorinated compounds, aldrin, dieldrin, endrin, chlordane, heptachlor, and toxaphene.

[b] Production of gamma isomer content in BHC was 17.1 million pounds in 1951, 10.7 million in 1955, and 6.9 million in 1960. Data in the table are on a gross basis.

[c] Figure not available, estimated to be about 8,800 thousand pounds.

SOURCES: U.S. Tariff Commission; U.S. Bureau of the Census; U.S. Bureau of Mines; communications from chemical industry.

Table 2. Domestic Disappearance at Producers' Level of Some Major
        Pesticidal Chemicals, United States (1,000 lb.)

Chemical	1950–51	1953–54	1958–59	1960–61	1961–62	1962–63
Aldrin-toxaphene group[a]	n.a.	n.a.	73,331	78,260	82,125	79,275
Benzene hexachloride[b]	9,600	7,610	4,276	4,577	2,404	1,299
DDT	72,688	45,117	78,682	64,068	67,245	61,165
Parathion	4,670	3,975	n.a.	n.a.	n.a.	n.a.
Calcium arsenate	39,583	3,190	n.a.	4,874	4,541	3,960
Copper sulfate[c]	122,449	74,054	84,230	78,220	80,815	80,599
Lead arsenate	30,174	16,000	n.a.	8,967	7,957	6,954
Pyrethrum (flowers)[d]	7,098	7,679	n.a.	n.a.	n.a.	n.a.
Rotenone (roots)[d]	7,027	6,428	4,827	3,888	3,598	3,336
2,4-D (acid equiv.)[e]	23,494	26,483	34,102	31,067	35,903	33,199
2,4,5-T (acid equiv.)[e]	2,822	3,877[f]	5,508	5,444	8,102	7,179

[a] Includes aldrin, chlordane, dieldrin, endrin, heptachlor, and toxaphene.
[b] Gamma isomer basis; includes lindane.
[c] Disappearance for all domestic uses, including industrial.
[d] Imports (preceding calendar years for 1958–1959 onward).
[e] Figures for 1950–51/1953–54 represent total disappearance at producer's level, export data not being recorded separately.
[f] Production for calendar year 1954, as reported by U.S. Tariff Commission.
SOURCES: U.S. Department of Agriculture, Agricultural Stabilization and Conservation Service, *The Pesticide Situation*; L. G. Arrington, *World Survey of Pest Control*, U.S. Department of Commerce, 1956 (as a secondary source of some data for earlier years).

herbicides used to control weed pests have increased more rapidly than either insecticides or fungicides.

According to the U.S. Tariff Commission, chlorinated hydrocarbons constituted more than 85 per cent of total insecticide and rodenticide sales by weight in 1961. Their sales continue to increase absolutely. Compared with 1957, the 1961 quantity of chlorinated hydrocarbons sold increased by 22 per cent. However, other classes of chemicals, principally the organophosphates and, to a lesser extent, the carbamates, have recently had an increasing share of insecticide and rodenticide sales.

DDT currently accounts for about one-half of total insecticide-rodenticide sales. Despite some problems of insect resistance, domestic disappearance of DDT continues slightly below the 1950–51 level (Table 2). It is so cheap and easy to apply that it can be expected to be in use for some time to come [3].

Production and sales of benzene hexachloride (BHC) increased very sharply in the late 1940's. In 1950, the production of BHC (gross basis) was only slightly less than that of DDT (Table 1). By 1963, however, production of BHC was only 6.7 million pounds, compared to 178.9 million pounds for DDT (Table 1). This decline in the production of BHC is undoubtedly the result of flavor changes in fruits and vegetables following use of this chemical [5]. Pronounced off-flavor in root vegetables and canning fruits and vegetables has been experienced, sometimes even when BHC was used on a nonfood crop grown earlier on the land. Off-flavor is especially noticeable when the food is processed by heat.

Areas Treated

Data showing the area in various land-use categories annually treated with insecticides were compiled in 1962 by the Entomological Society of America. A summary is set out in Table 3.

Approximately 5 per cent of the land area of the United States, excluding Alaska and Hawaii, is treated with insecticides annually. Cropland and cropland pasture constitute more than three-fourths of the treated area, with cereal crops (principally corn and small grains) accounting for nearly one-half of the treated area in this category. According to these data, 15 per cent of the land devoted to grain production was treated with insecticides, with the average rate of application being one pound of active ingredients per acre (Table 3). An estimated 80 per cent of the area of fruit and nut planting receives annual insecticide application. The average quantity of insecticidal chemicals applied on the 2.3 million acres of orchards treated is estimated at 7 pounds per acre annually. This very high rate of usage is

Table 3.   Acreages Treated Annually with Insecticides, United States
(Excluding Hawaii and Alaska), 1962

Land use	Acres in category (million)	Acreage on which insecticides applied (million)	Percentage of category treated
Forest lands	640	1.8	0.28
Grassland pasture	630	1.6	0.25
Desert, swamps, dunes, and wildland	77	2.5	3.24
Water areas	32.6	0	—
Cropland and cropland pasture	457	68.6	15.0
Of which			
Fruits, nuts	2.8	2.3	80
Cotton	15.8	11.9	75
Vegetables	4.1	2.1	50
Grains	216.6	32.5	15
All other crops, etc.	217.6	19.9	9
Urban or built-up areas	53	15	28.3
Nonforested parks, wildlife refuges, duck reserves, national defense sites	43	—	—
Total U.S. acreage (48 states)	1,934.6	89.5	4.6

SOURCE: U.S. Department of the Interior, Fish and Wildlife Service, *Pesticide-Wildlife Studies,* Circular 167, June 1963.

exceeded, however, by that for cotton [6]. Table 3 shows that an estimated 75 per cent of the 11.9 million acres in cotton receives insecticide applications. This is at an average rate of 7.5 pounds of insecticide per acre [6]. Over the whole area of 89.5 million acres annually treated with insecticides, the average amount of ingredients estimated to be applied per acre is 2.5 pounds [6].

The area treated with herbicides in 1962 is estimated by the U.S. Department of Agriculture at more than 85 million acres, representing an increase of more than 60 per cent since 1959 [5]. Since there is considerable overlapping of areas treated with insecticides and herbicides, it has been estimated that approximately 1 acre in 12 of the nation's total land area receives one or more applications of a pesticide each year [7]. On the other hand it is estimated that about 75 per cent of the land area

Table 4. Extent of Pest Control and Defoliation Treatment, Regions and United States, 1952 and 1958

| | 1952 | | 1958 | | |
Region	Acreage treated	Average times treated	Acreage treated	Average times treated[b]	Total acreage treated[c]
	1,000 acres	Number	1,000 acres	Number	1,000 acres
		Total treatment[a]			
48 states	60,103	1.94	96,109	1.65	158,440
Northeast	3,204	3.26	4,768	2.52	12,022
Lake states	4,165	1.64	9,238	1.35	12,474
Corn belt	7,939	1.25	19,230	1.12	21,626
Northern plains	7,629	1.07	17,613	1.10	19,438
Appalachian	3,613	2.56	3,735	2.12	7,926
Southeast	4,764	3.40	3,769	3.21	12,097
Delta states	4,885	3.38	4,798	3.38	16,229
Southern plains	7,493	1.77	10,052	2.21	22,261
Mountain	7,213	1.36	12,083	1.28	15,437
Pacific	9,198	1.77	10,823	1.75	18,930
		Insect and disease control			
48 states	29,002	2.86	37,234	2.61	97,199
Northeast	1,728	5.13	2,058	4.45	9,163
Lake states	743	4.22	1,365	3.10	4,224
Corn belt	1,604	1.96	4,199	1.54	6,458
Northern plains	702	1.44	4,573	1.35	6,169
Appalachian	2,810	2.97	2,275	2.79	6,353
Southeast	4,493	3.53	2,893	3.87	11,209
Delta states	4,241	3.71	2,611	5.27	13,770
Southern plains	5,810	1.98	6,643	2.78	18,473
Mountain	2,341	2.03	5,720	1.56	8,954
Pacific	4,530	2.35	4,897	2.54	12,426
		Weed and brush control			
48 states	31,101	1.08	55,222	1.03	56,865
Northeast	1,476	1.07	2,597	1.05	2,721
Lake states	3,422	1.07	7,834	1.05	8,211
Corn belt	6,335	1.08	15,026	1.01	15,163
Northern plains	6,927	1.03	13,031	1.02	13,259
Appalachian	803	1.11	1,433	1.07	1,540
Southeast	271	1.37	791	1.01	803
Delta states	644	1.16	1,342	1.04	1,395
Southern plains	1,683	1.04	1,900	1.07	2,029
Mountain	4,872	1.04	6,066	1.01	6,156
Pacific	4,668	1.19	5,202	1.07	5,588

[a] Includes defoliation.

[b] Total acreage divided by acreage treated. May not check due to rounding.

[c] Sums of state data. Acreage times number of treatments.

Source: U.S. Department of Agriculture, Economic Research Service, Farm Economics Division (Paul E. Strickler and William C. Hinson), *Extent of Spraying and Dusting on Farms, 1958 with Comparisons,* Statistical Bulletin No. 314, May 1962, p. 10.

of the 48 conterminous states has *never* been treated with insecticides.

A survey of the extent of spraying and dusting on farms in 1958 was carried out by Strickler and Hinson [8]. They estimated the area subjected to pest control treatments and the average number of treatments in 1958 for each of the 48 mainland states and for major crops, based on 20,500 mailed questionnaires completed by voluntary correspondents of the Statistical Reporting Service. These estimates are compared with the results of another survey conducted by the U. S. Department of Agriculture for the year 1952 and, in the case of cotton, also with figures for the year 1949.

Table 4 shows the total areas treated with pesticides and defoliants for the United States and each region and also the areas treated for insect and disease control. The total area treated increased from 60 million acres in 1952 to 96 million acres in 1958. The largest increases took place in the corn belt, the northern plains, and the lake states. Most of the increases in these areas can be attributed to increases in acres treated for weed and brush control. By contrast, in the southeast and Delta states, the total acres treated and the acres treated for control of insects and disease declined over the same period.

Table 5 presents Strickler and Hinson's estimates of areas and numbers of treatments for insect and disease control, weed and brush control, and defoliation by crops for the years 1952 and 1958. These data show an over-all increase of 28 per cent in the total area of land treated for insect and disease control and a number of shifts in the relative importance of certain crops in this increase is interesting. In 1952, very little corn was treated for insect and disease control, but by 1958 the acreage of corn treated was larger than any other crop except cotton, alfalfa, and clover. Even though the treatment of corn acreages is estimated to be slightly over one treatment per year, requiring much less poisonous material than an acre of cotton or other high value crops, the expanding acreage of corn treated with insecticide implies the exposure of a larger part of the environment to insecticide materials. Cotton acreage, treated once over, declined over this

Table 5. Extent of Pest Control and Defoliation Treatment, by Crops
United States, 1952 and 1958

	1952		1958		
Crop or land	Acreage treated	Average times treated	Acreage treated	Average times treated	Total acreage treated[a]
	1,000 acres	Number	1,000 acres	Number	1,000 acres
	Insect and disease control				
Total or average	29,002	2.86	37,234	2.61	97,199
Cotton	13,066	3.06	8,144	4.41	35,943
Alfalfa and clover	3,046	1.27	6,639	1.34	8,912
Corn	414	1.08	4,519	1.13	5,109
Fruits and tree nuts	3,459	4.55	3,516	4.96	17,436
Vegetables	2,270	3.25	2,946	3.12	9,201
Potatoes	1,071	5.12	1,193	4.52	5,393
Tobacco	1,407	2.92	862	2.64	2,272
All other crops and land	4,269	1.42	9,415	1.37	12,933
	Weed control				
Total or average	31,101	1.08	55,222	1.03	56,865
Small grains	17,107	1.04	24,853	1.00	24,921
Corn	9,173	1.05	21,599	1.03	22,136
Pasture and rangeland	2,192	1.14	3,427	1.10	3,765
Cotton	[b]	—	810	1.11	902
All other crops and land	2,629	1.37	4,533	1.13	5,141
	Defoliation				
Total or average	[c]	—	3,653	1.20	4,376
Cotton	[c]	—	3,175	1.22	3,860
All other crops and land	[c]	—	478	1.08	516

[a] Sums of state data. Number of acres times number of treatments.
[b] Included with all other crops and land.
[c] Not available.
SOURCE: U.S. Department of Agriculture, Economic Research Service, Farm Economics Division (Paul E. Strickler and William C. Hinson), *Extent of Spraying and Dusting on Farms, 1958 with Comparisons,* Statistical Bulletin 314, May 1962, p. 9.

Table 6. Estimated Extent and Cost of Chemical Weed Control, by Crops, United States, 1959

Crop	Acreage treated		Combined cost of herbicide and application[a]	
	Pre-emergence	Post-emergence	Pre-emergence	Post-emergence
	1,000 acres	1,000 acres	$1,000	$1,000
Corn	2,235.4	17,816.5	8,226.3	29,753.6
Cotton	1,000.6	552.7	3,221.9	1,406.1
Soybeans	545.5	10.0	2,296.6	17.5
Small grains	0	20,723.3	0	37,094.7
Rice	0	502.0	0	888.5
Peanuts	31.9	3.0	107.2	9.0
Sugar beets	81.6	42.6	427.6	197.2
Sorghum	8.0	2,085.0	48.0	6,463.5
Forage seeds	0	281.8	0	1,868.3
Vegetables	71.9	204.1	582.1	835.6
Fruits and nuts	0	5.4	0	42.8
Strawberries	2.0	3.3	35.2	20.4
Ornamentals	.2	2.2	2.0	43.1
Lawns	2.6	57.3	680.0	809.1
Hay	0	272.4	0	1,691.6
Pastures	30.1	2,370.0	30.1	5,759.1
Rangeland	0	2,011.0	0	6,173.8
Noncrop land	27.2	1,943.5	2,596.2	17,141.7
Totals	4,037.0	48,886.1	18,253.2	110,215.6

[a] Calculated from average costs incurred by farmers and other land owners in the states reporting.

SOURCE: U.S. Department of Agriculture and U.S. Federal Extension Service, *A Survey of Extent and Cost of Weed Control and Specific Weed Problems,* ARS 34–23, March 1962.

period because of large reductions in acreages planted to cotton. The same can be said about the decline in the acreage of tobacco. While the acreage of fruit and tree nuts treated showed little change between 1952 and 1958, the area of pasture, rangeland, small grains, and sorgham (grouped in "all other crops and land" category) more than doubled during this period. Most of this increase took place in the mountain and northern plains regions where grasslands and small grains were treated extensively for grasshoppers.

A more detailed breakdown by crops is available for estimated areas treated with herbicides in 1959 from a survey by the U.S. Department of Agriculture. Table 6 shows the acreage treated with pre-emergence and post-emergence weedkillers and the combined costs of herbicide and application as estimated from this survey. These estimates show 52.9 million acres as the combined acreage subjected to pre- and post-emergence herbicides. Since it is likely that many acres were subjected to both types, the actual acreage treated is an area smaller than 52.9 million acres. Although these data show corn and small grain growers to be the most important users of herbicides on an area basis, the farmer clientele for herbicides was very widely distributed over an array of crops.

In 1959, chemical methods of weed control were used on about 13 per cent of the total harvested acreage of agronomic crops. Of the remaining 87 per cent, weeds were controlled by cultural, ecological, mechanical, and other nonchemical means. The movement to chemical weed control has derived momentum in recent years from the increasing costs and diminishing availability of casual rural labor. It is also a necessary adjunct to the trend toward larger-scale farming that is characterized by specialization and modern machinery—both requirements for economic operations. It is likely, therefore, that the use of herbicides will continue to expand. Further increases may also be expected in non-agricultural uses of herbicides, including the prevention of vegetative growth or control of undesired species on median and border strips of highways and clearings for power transmission lines.

Spillover Hazards

Do these trends in the kinds and quantities of pesticides used and the expansion of the area to which they are applied give rise to any hazards? If so, what are the nature of these hazards? Are any economic problems posed for society as a consequence?

Past discussion of the pesticide problem by others indicates that there are hazards associated with these trends. Some of the hazards are real, while others are still conjectural. Many of these hazards take the form of adverse side effects of pesticide use or conflicts with other objectives. If these side effects have their incidence on persons other than those who make the decisions governing pesticide use, economic problems emerge in the form of cost–benefit distribution and resource allocation problems.

If pesticide use does in fact result in costs being borne by persons who do not share in the benefits, to the extent that too many resources are being allocated to pesticide production and use, relative to ends desired, then the economist is compelled to subject such problems to economic analysis. It should be noted that economists are equally as concerned about spillover rewards where persons not sharing the costs or the decision-making responsibilities receive unearned rewards. This latter situation can result in too few resources being allocated to a particular activity, and causes a divergence from optimum results.

In this section we are concerned with the spillover hazards from pesticides that result in less than optimum use of resources. The fact that spillovers exist does not always reflect poor decisions by individuals or groups. There are a number of reasons why such spillovers, or external diseconomies as they are often termed, may exist. These will be treated in the following discussion.

External diseconomies refer to costs imposed by the actions of a decision-making unit upon other individuals, firms, or communities. Water and atmospheric pollution provide notable examples. Firms which adopt a least-cost method of disposing of their effluents by discharging them into a river may impose costs, including the costs of curtailed opportunities, upon others downstream. In consequence, the total outcomes of individual decisions in the river basin may fall considerably short of the optimum outcome, viewed on a basin-wide or community basis [9].

External economies (spillover rewards) may also result in failure of the market system to achieve the socially optimal outcome. If a large city develops a mass transit system to transport persons to the central city by means of a subway, then benefits accrue not only to those who ride the trains, but also to those who use the streets after the system is activated. That is, the less congested streets result in such improvements as better traffic movement, less air pollution, fewer accidents, and a more effective police and fire department. In this situation, a strong argument can be made for general support of the transit system through taxation since benefits do accrue to nonriders. In fact, if such support is not provided, it can lead to less than optimum investment in mass transit and many benefits could be lost.

In many areas of economic activity this kind of divergence between private and social returns is small or nonexistent. The competitive market system can be accepted, in many cases, as yielding an outcome which is reasonably close to the optimum. Alternatively, such imperfections as do occur may be preferred to the implications of introducing social controls or adopting an alternative to the market as a method of allocating resources and distributing costs and benefits. There are, however, some fields of economic activity in which divergences of significant magnitude appear between private and social costs and benefits.

Indivisibilities in investments may lead to suboptimal use of resources from a social viewpoint. Many inputs can be used only on a scale which is beyond the resources of a single firm or individual. A common textbook example is mosquito control. It is true that antimalarial mosquito control in small communities has sometimes been undertaken by single firms and has paid off in increased labor productivity and reduced absenteeism. Generally, however, the fact that the expected benefits to the individual firm are less than the total costs of the program serves to prevent investments that might result in total benefits to all parties far in excess of total costs. This is, of course, the reason that it is often necessary for health and education investments to be undertaken by government.

Both technological externalities and indivisibilities may, therefore, bring about a situation in which the market fails to establish appropriate values and thus fails to allocate resources in the manner required to maximize social welfare. The standard means of dealing with technological external diseconomies and economies is to enlarge the decision-making unit to correspond to the effects [10]. Spillovers are then eliminated by rendering all effects internal.

Pest control offers excellent examples of both technological externalities and indivisibilities. The nature and incidence of these are examined in subsequent chapters. Some preliminary consideration of these imperfections, however, will illustrate their potential importance as a factor leading to divergence between private and social costs and will help to grasp the nature of the economics embodied in pesticide policy.

The adoption in many countries of public programs affecting pesticide usage reflects the presumed need for control over external effects of the chemicals employed. One factor capable of producing technological externalities is drift. Chemical sprays often fall partly outside the target area. Deposition of chemicals on unintended areas may cause costs on other farms, affecting crops and livestock. As a result of drift, reduced crop yields due to phytotoxicity of the chemical agents or reduction of market quality by contamination may occur. Affected forage crops fed to livestock may be rendered unmarketable because of the likelihood of excessive residues in milk or meat. Drift, of course, can bestow benefits on others by providing free control of pests on neighboring tracts.

Pollution of streams, lakes, and underground water by unintended or careless deposition of chemicals on water surfaces or by runoff or seepage has occurred. Further examples of spillovers are kills of beneficial insects such as bees, or damage to fish and wildlife populations. Any effects of toxic residues on human health and uncompensated or inadequately compensated occupational diseases suffered by workers exposed to chemicals would also represent instances of disparity between private and social costs.

Certain economic and technical relationships in agriculture may also predispose the situation towards spillovers. One hypothesis [11] suggests that the typical situation facing the farmer is as illustrated in Figure 1. The dosage–mortality, or dosage–yield relationship, flattens out so that, beyond a certain level, very little additional insect kill or yield gain is achieved by using additional chemical. At the same time, the slope of the price line relating the price of a unit of agricultural output to the price of a unit of

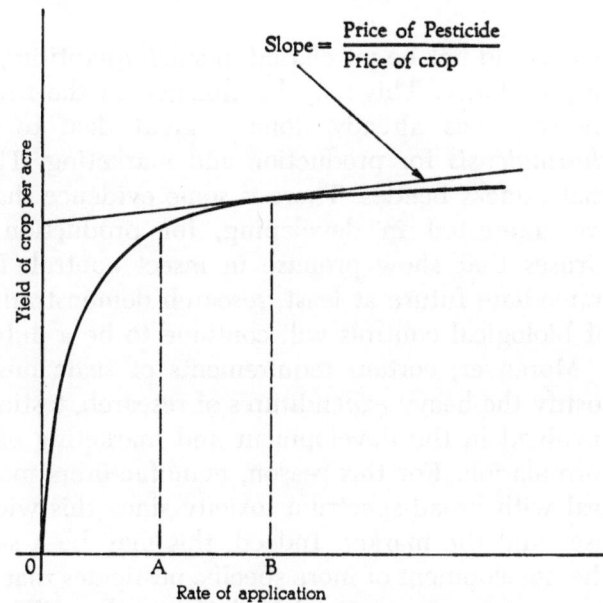

Figure 1. Relationship between yield and rate of pesticide application (hypothetical).

pesticide application is believed to be typically flat. In other words, pesticide chemicals and their attendant application costs required to control pests are very cheap in relation to the unit value of crop or livestock products.

In these circumstances, the optimal quantity of pesticides from the individual farmer's point of view is *OB* and may be substantially in excess of the socially optimal level, taken for illustration to be *OA*. Massive doses for very small additional gains may, therefore, result. More will be said about this in Chapter 2.

While everyone would not agree that pesticide costs are generally so low in relation to the value of the product, the relationships depicted undoubtedly have relevance in a number of situations, particularly with high value crops such as fruit and vegetables.

Indivisibilities are present in the field of biological pest control. Thus, with a few exceptions such as the sale of Australian ladybug beetles and a few insect pathogens, biological control in the past could not be purchased in small quantities, but was a public responsibility. This may be changed in the near future. Private industry has already done a great deal to develop *Bacillus thuringiensis* for production and marketing. This is a pathogen that attacks beetles. There is some evidence that industrial firms are interested in developing, for production and marketing, viruses that show promise in insect control. However, for the immediate future at least, research demonstrating the usefulness of biological controls will continue to be a public responsibility.

Moreover, certain requirements of scale must be fulfilled to justify the heavy expenditures of research, testing, and promotion involved in the development and marketing of a new chemical formulation. For this reason, manufacturers may prefer a chemical with broad-spectrum toxicity since this widens the range of uses and the market. Indeed, this may be a serious obstacle to the development of more specific pesticides that are advocated by some observers as a partial solution to the spillover problem. With scale requirements as they are, some limitations of competition may easily arise and the expectation of low rates of return resulting from entry of another pesticide into a limited market may tend to perpetuate the use of multiple-purpose chemical agents. Such a situation could lead to deleterious social consequences, and the development of products less subject to technological externalities may be discouraged.

The hazards of increasing use of pesticides in both a quantity and an area dimension have been conceptualized as hazards of spillovers in which actions influence the productive capacity of resources or threaten the achievement of objectives beyond the decision-making unit. A solution to the problem does not necessarily involve getting rid of all technological external diseconomies. The adoption of this alternative could result in a serious net reduction of social welfare such as, for example, banning all chemicals capable of creating external diseconomies. The need is, rather, for a program providing the optimal outcome in terms of agricultural output and resource saving, public health, recreation, and aesthetic values. Such a program may well involve compromise between conflicting objectives and safeguarding against some spillovers, but permitting others in accordance with the balance to society.

REFERENCES

1. E. F. Knipling, "Alternative Methods in Pest Control," *Symposium on New Developments and Problems in the Use of Pesticides,* Publication 1082 (Washington, D.C.: National Academy of Sciences-National Research Council, 1963), pp. 23–38.
2. G. C. Decker, "Insecticides as a Part of the 20th Century Environment," Ecological Society of America (Bloomington, Indiana), August 25, 1958, p. 11. Mimeo.
3. Stanley A. Hall, "Trends in Insect Control Agents," *Journal of the Washington Academy of Sciences,* Vol. 52, No. 2, February 1962, pp. 29–35.
4. U.S. Department of Agriculture, Agricultural Stabilization and Conservation Service, *The Pesticide Situation for 1962–63,* September 1963.
5. C. H. Mahoney, "Flavor and Quality Changes in Fruits and Vegetables in the United States Caused by Application of Pesticide Chemicals," *Residue Reviews,* Vol. 1, 1962, pp. 11–23.
6. U.S. Department of the Interior, Fish and Wildlife Service, *Pesticide-Wildlife Studies,* Circular 167, June 1963, p. 3.
7. *Ibid.,* pp. 1–10.
8. U.S. Department of Agriculture, Economic Research Service (Paul E. Strickler and William C. Hinson), *Extent of Spraying and Dusting on Farms, 1958, with Comparisons,* Statistical Bulletin No. 314, May 1962, p. 28.
9. Allen V. Kneese, *The Economics of Regional Water Quality Management* (Baltimore: The Johns Hopkins Press, 1964), pp. 38–53.
10. J. W. Milliman, "Can People Be Trusted with Natural Resources?" *Journal of Land Economics,* Vol. 38, No. 3, August 1962, pp. 199–218.
11. P. M. Hillebrandt, "The Economic Theory of the Use of Pesticides. Part I," *Journal of Agricultural Economics,* Vol. 13, No. 4, January 1960, pp. 464–72.

2

A Decision Framework, Alternative Actions, and Values

The solution to the "pesticide problem" lies in the selection of those courses of public action that will yield the best outcomes for society as a whole—taking into consideration the values placed upon the multiple objectives concerned. In Chapter 1 the problem was defined as one of spillovers (external economies and external diseconomies) and technological indivisibilities resulting from a growing usage in quantity and the expanding array of uses of chemical pesticides.

Policy in the area of pest control should be concerned with the improvement in satisfaction resulting from the use of the nation's resources. Therefore, the decisions upon which policies are based need to be concerned with alternative actions, potential consequences accompanied by their degree of certainty, and the values placed on the consequences.

It is the purpose of this chapter to develop a framework, useful for decisions, that shows the kinds of information needed for policy decisions, and to show the relation among variables in the problem as a basis for developing criteria with which to judge alternative policies. As a result, such a discussion may be expected to help in assessing the needs for further research and in the selection of means of analyzing the problem.

THE FRAMEWORK

A framework representing the effects of pesticide usage and

containing criteria for optimal usage can be developed, using the concept of a production function (input–output relation) for the pesticide-using part of the economy in conjunction with a function-relating satisfaction derived from that production.

In very general terms, we can say that the pesticide users, principally agriculture and forestry, produce goods that are of positive benefit to society. They also produce goods that are of negative benefit to society in the form of any adverse effects resulting from the use of pesticides. (See the section on spillover hazards in Chapter 1.) These negative benefits can be treated as a cost to society. Society's objective can then be stated simply as securing that level of pesticide usage, given the technology at any point in time, that provides the maximum positive benefits over and above the negative benefits or costs associated with that level of usage. In this way net satisfaction or net social value is maximized. If a time dimension is introduced, the social objective becomes the maximization of the present value of a discounted stream of annual net social values over a period of years for which a particular pest control policy and technology are being evaluated.

Let us be a little more rigorous with this definition of the optimal level of pesticide use. The positive benefit from the last increment of pesticide use is defined as V which includes the direct and external benefits from pesticide use. In addition the cost of the last increment of pesticide use is defined as $C_p + P$. C_p is the cost of the material in the market, plus application costs, and can be considered to be the social value of the resources used in the development, manufacture, sale, and application of the chemical. P is a measure of any external adverse effects such as damage to bees, pest predators, wildlife, etc., a negative benefit from the last increment of pesticide applied. P may be greater than, or equal to, zero.

Now, if V is greater than $C_p + P$, not enough pesticides are being used. If V is less than $C_p + P$, too much pesticide is being used. However, if V equals $C_p + P$ then the optimal amount of pesticides are being used and net social value is at a maximum. This is true if, and only if, the levels of other inputs in the production process have been held constant. In other words, given the levels of land, labor, and other capital inputs, the positive benefits resulting from the use of one more increment of pesticide chemical must not be less than the cost of the resources used to develop, manufacture, sell, and apply that increment of pesticide. Otherwise, the net social value would be reduced.

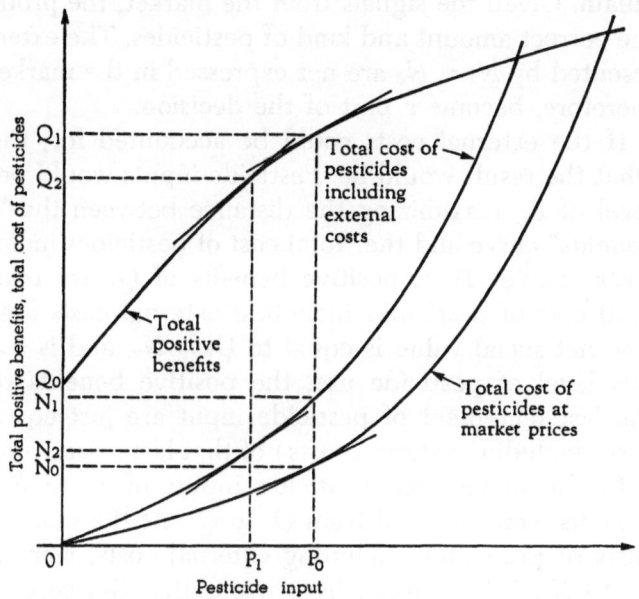

Figure 2. Optimum use of pesticides considering external costs.

Figure 2 provides a graphic expression of the maximization of the net social value just described. The figure portrays a hypothetical relationship where positive benefits equal to OQ are produced with no pesticides. As pesticides are added to a level of P_0, other inputs held constant, positive benefits accrue to a level of Q_1. At this point, the distance between the "total positive benefits" curve and the "total cost of pesticides at market prices" curve is a maximum and equal to $Q_1 - N_0$. However, if there are external costs associated with this level of pesticide usage, then the real costs of the pesticide input P_0 is N_1 and the net social gain is really $Q_1 - N_1$. It is obvious from Figure 2 that $Q_1 - N_1$ is less than $Q_1 - N_0$ and the value of the positive benefits from the last increment of pesticides does not exceed the total cost (including external costs) of that last increment. More pesticides are applied than is socially optimal. It should be pointed out that this can occur without affixing blame, or the labeling of any group as the villain. Given the signals from the market, the producers applied the correct amount and kind of pesticides. The external costs represented by $N_1 - N_0$ are not expressed in the market and cannot, therefore, become a part of the decision.

If the external costs could be accounted for, Figure 2 shows what the result would be. Pesticide inputs would be reduced to a level of P_1, maximizing the distance between the "total positive benefits" curve and the "total cost of pesticides including external costs" curve. Total positive benefits of Q_2 are realized and the total cost of pesticides including external costs are equal to N_2. The net social value is equal to $Q_2 - N_2$, and is a maximum. At this level of pesticide use, the positive benefits resulting from the last increment of pesticide input are just equal to the total cost (including external costs) of that last increment of pesticides.

In the move from pesticide inputs of P_0 to P_1, the positive benefits were reduced from Q_1 to Q_2. At the same time, the total costs of pesticides, including external costs, were reduced from N_1 to N_2. It is graphically obvious that the costs were reduced more than the benefits by this action. That is to say, that $Q_1 - Q_2$ is less than $N_1 - N_2$.

This framework shows what economists mean when they refer to the external costs of pollution, and it shows how external costs, when ignored, can result in resource allocation which is less than optimal. This framework is concerned only with the attainment of maximum net social value or what might be called efficiency. It does not deal with the questions of distribution of the gains. It is possible that the gain has come at the expense of the welfare of others involved in the pest control question. If such is the case, policies will need to include a means whereby those who gain can compensate those who lose.

Implications of the Framework

The controversy surrounding the pesticide question is based on the lack of knowledge in four areas within this framework. First, the positive benefits resulting from incremental changes in the output of agriculture and forestry are not certain (the slope and shape of the total positive benefit function in Figure 2). Second, the magnitude of the external costs of pesticide usage are not known, nor is the behavior of these external costs resulting from incremental increases in usage known. Third, a means of placing values on the physical external costs is not yet developed. Fourth, the substitution rates of other inputs such as land, labor, and fertilizer for pesticides are not known.

Knowledge concerning the substitution rates between pesticides and other inputs used in agricultural production requires some explanation. If pesticides can be substituted for land, labor, or fertilizer, then the cost of replacing a part or all of currently used materials can be computed with the knowledge of these substitution ratios. Substitution would increase the demand for the resource being substituted and result in an increase in the price of the resource if an attempt is made to maintain the level of output. For instance, if a pesticide cannot be substituted for fertilizer, but is a technical complement of fertilizer, then the levels of use of the two resources are tied together. Consequently, an increase in pesticides will be associated with an increase in fertilizer use and a decrease in pesticide use will be associated with a decrease in fertilizer use. It is possible that this bundle of resources, viz., pesticides and fertilizer, will substitute for land and/or labor and, if this is the case, any cutback in use of pesticides would result in a cutback in fertilizer use and an increase in land and labor until the cost of the additional land and labor could not be covered by the additional production attributed to it.

We have hypothesized that the use of pesticides is subject to costs with two components. One is the market price representing the social value of the resources used to produce pesticides. This we called C_p. The other is the external cost representing the adverse effects of pesticides which may be imposed upon others. This cost was defined as P.

If P is greater than zero, the policy avenues that are available to deal with these external costs fall under four general headings presented below.

1. If the same level of positive benefits from pesticide use can be obtained by the use of other pest control inputs that are relatively free of external costs, then substitution of these other inputs can be encouraged by administratively raising the price of pesticides relative to other pest control inputs.

2. If the positive incremental benefits from pesticide use are estimated to be less than the combined incremental market and external cost resulting from the use of pesticides, and if substitution of inputs is not economically possible, then the net social value can be increased by banning the use of the pesticide or pesticides in question except in cases where use is considered essential and where very careful controls can be exercised over the use.

3. If, through research, alternative modes of pest control can be developed which will reduce the external costs, and if the capitalized cost of research is less than the estimate of the external costs eliminated, then such research can be encouraged by policy. This approach involves both altering the nature of chemical pesticides and the substitution of other forms of control for all or part of the chemicals.

4. If the external costs arising from the use of pesticides are primarily the result of ignorance with respect to existing alternatives or ignorance with respect to the proper use of pesticides capable of adverse spillover effects, then policy can develop educational and regulatory programs to insure that potential pesticide users are informed on the types of control available and their proper use, and that proper use is effected.

ALTERNATIVE CONTROL METHODS

Before any administrative actions are taken to alter the relative price of pesticides or to prohibit the use of certain materials, the existing alternatives to the use of chemicals with adverse side effects should be examined. Since public policy for pest control will probably not flow from one enormous decision model, but rather from a series of decisions on specific problems, the following review of alternative pest control actions is presented.

Preventing Introduction and Spread of Pests

The first line of defense is to prevent the introduction and establishment of new insect pests and weeds from other countries. Of several hundred insect species that are particularly destructive and require some measure of control in the United States, no less than 123 are immigrants [1]. Quarantine regulations controlling the import and interstate movement of products which harbor insect pests are used to prevent the entrance of new pests and the spread of pests to new areas inside the country. The costs of such activity are borne publicly by the agencies concerned with immigration, international trade, and interstate commerce.

Prompt measures to suppress newly established pests are, of course, the second line of defense. Early pest detection, identification, and eradication can do much to obviate the necessity for widespread pesticide treatments which may become necessary when the new species become distributed over an extensive area. Such eradication campaigns in the past have involved fairly heavy rates of pesticide application with the possibility of creating spillovers of a drastic nature, but confined to a limited area. Costs of such campaigns are borne primarily by the U.S. Department of Agriculture and its agencies concerned with pest control.

Cultural Practices and Sanitation

Cultural methods, involving such practices as the destruction of crop refuse, deep plowing, mechanical cultivation or clipping of weeds, rotation of crops, and the timing of planting to avoid or minimize exposure to pests, are the oldest forms of pest control. Despite a few notable successes, such as the use in south Texas of deadlines for planting cotton and for destruction of stalks after the harvest, this group of methods is of limited effectiveness and, while important in an over-all plan of pest control, they cannot usually be considered as more than a useful supplement to pesticides [2].

Costs of this class of alternatives are borne by individual property owners and to some extent by government, where such practices are followed on public lands. The magnitude of the costs is variable depending upon the practice and the land area involved.

Use of Natural Enemies of Pests

An important group of alternatives consists of the various biological control methods. Included in this broad class are such methods as the introduction and establishment of predators, parasites, and pathogens of either insects or weeds and the breeding of resistant strains of plants and animals. The use of natural enemies boasts of some spectacular successes. However, some who have urged almost complete reliance on this approach have not recognized that several important limitations prevent its general application to all major pests. On the other hand, limited success in the past does not necessarily imply that further work on nonchemical control methods offers little prospect of success on a broader front.

More than 650 species of insect parasites and predators have been collected and introduced to the United States. At least 100 have become established and it is claimed that these are "giving outstanding control of twenty of our serious insect pests" [1]. Successes with control of noxious weeds have also been obtained. Introduction of a beetle brought Klamath weed under control in the western states. A number of successes have also been recorded overseas.

It was estimated in 1954 that at least $500,000 was expended annually in the biological control of insect pests, including the search for beneficial insects in foreign areas and their importation [3]. Concern with evidence of environmental pollution from chemicals has served to enlarge this search. In 1963, it was estimated that approximately one-half of the research effort on conventional insecticides had been shifted to research on non-chemical control methods or on chemicals which are specific in their targets [4].

A number of limitations of natural enemies of pests have been enumerated by Van Den Bosch and Stern [5] and by the Agricultural Research Service [1]. Many insects appear to be completely free of parasites or do not have important natural enemies capable of effective control. More than one parasite or predator may be needed or an alternate host may be needed if the parasite is to persist; otherwise the rate of increase of predators and parasites may not be sufficient to control a sudden outbreak.

Few of the biological control alternatives destroy the pest completely. Less than complete control may be insufficient to protect high value crops or to meet exacting standards imposed by the market and/or pure food laws.

It has been pointed out that overdemanding pure food laws and market standards may cause growers to resort to heavy treatments with broadly toxic pesticides [5]. Many argue that these pesticide treatments may pose greater hazards to humans than a few thrip tarsi or aphid antennae.

The costs of control by natural enemies have, in the past, been largely borne by government or by producer groups. Since introduction of a natural enemy requires ecological studies beforehand, the costs can be quite large—usually beyond the resources of individual producers. This set of alternatives also exhibits the indivisibilities discussed earlier.

Resistant Plant Varieties

The breeding of resistant plants is a well-known technique in disease control. The method has also been applied with outstanding success to the control of certain insects and nematodes. Wheat varieties resistant to Hessian fly and alfalfa resistant to spotted alfalfa aphid are examples. Some varieties withstand heavy infestations, while others exert a profound influence on insect fecundity [1].

Developing resistant plant varieties may frequently be costly and time-consuming. These costs have historically been borne by universities, government agencies, and seed companies. The time required precludes the use of this alternative to control sudden outbreaks of pests.

Pre-emptive Sterile Insects

A more recent alternative offers a new dimension in pest control methods [4]. This is the use of sterile insects, in large numbers, to prevent reproduction. This alternative was applied in the elimination of the screwworm in the southeast United States. It was also successful in eliminating melon fly in tests on the remote island of Rota. Pilot studies are under way with codling moth, boll weevil, and several species of fruit fly.

The requirements of this alternative are as follows: (a) a method of mass rearing or sterilization, (b) a method of producing sterility and retaining reproductive urge, and (c) a low level in the population at some stage of the seasonal cycle [2].

A number of chemosterilants have been found which, used in conjunction with synthetic attractants, may enable the sterilization of insects in the field and reduce the need for costly facilities for breeding and population increase. However, we should not be overly optimistic about the use of chemosterilants [6]. Only a limited number of effective compounds, all ethyleneamine derivatives, are known at present. As with most control methods involving a biological component, chemosterilants cannot be used by individuals as ordinary agricultural chemicals are used. They must be applied over wide areas and under carefully controlled conditions. However, the sterile insect method becomes more efficient as the population declines. This is in marked contrast to insecticides or pathogens where the cost of eliminating the last 1 per cent of an insect population is very high.

Alternative Ways of Using Pesticides

Finally, of course, there are many alternative actions open in the use of pesticides. Hazards can be modified by controls over pesticide use, by research to develop chemical agents with more desirable properties, by education of users, regulation of residues on foodstuff, etc.

In detail, outcomes in agriculture, forestry, public health programs, and fish and wildlife management can be influenced by alternatives such as registration, labeling [7], education of users, licensing of operators and enforced custom application of certain chemicals, supervision of commercial applicators [8], residue control, and use of taxes and subsidies to discourage and promote certain chemicals.

VALUES AND EXTERNALITIES

Earlier, it was noted that one of the problem areas in the development of pesticide policy concerns the valuation of the external effects resulting from pesticide use. While there are external benefits resulting from pesticide use, such as the control achieved by agricultural producers because of their neighbor's use of pesticides and the reduction of insect pests damaging to wildlife, most of the controversy centers around such external costs as hazards to human health, damage to fish and wildlife, and the alleged development of an ecological imbalance.

Our decision framework is based on the assumption that values can be placed on the external effects, and that they are known. Yet, as we have stated previously, many of these external effects do not have values that are explicitly determined in the market.

It is perhaps an understatement to say that assigning values to such things as illness, a human life, a bird, a fish, etc., is a complex problem. Yet the valuation problem is inescapable in decision making. There are those who express a strong distaste for value judgments. However, the act of making a decision implies the placing of values on alternative outcomes.

Since decisions concerning pesticide policy involve multiple objectives, it is desirable to translate consequences into some common measure. Where consequences do not lend themselves to quantification or have no universal money value (such as a price determined in an established market), certain consequences that are desired or that must be prevented can be incorporated into our decision framework in the form of constraints. In this way, the quantifiable sacrifices in other values that are required can be used as a measure of the cost of the particular consequence desired or to be prevented.

In effect, placing values on the adverse effects of pesticide use amounts to a determination of the costs of eliminating the effect of development of an alternative, or of the sacrifice in positive measurable benefits by decreasing the level of use. If the decision is made to develop the alternative or restrict the use, the additional cost of the alternative or the sacrifice in positive benefits represents an implicit estimate of the value of the adverse effect. We do not know how much more would be sacrificed or how small the sacrifice must be to change the decision. But we do have a value on which action was based.

PROBABILITIES

In our decision framework, the assumption has been made that all of the parameters, if known at all, are known with certainty. This abstraction from reality allows the analysis to remain uncluttered and free of probability distributions. However, reality must be faced. In so doing, it is found that few of the parameters are single-valued.

To begin with, the output response from changes in inputs can vary widely because of weather, insects, disease, and other exogenous forces. For example, the productivity of pesticides is a function of the insect or weed infestation, which is itself some function of weather and diseases. Each of these variables is subject to a probability distribution and, therefore, the outcome becomes probabilistic.

Many of the adverse effects attributed to pesticides are classified as potential effects. This means that logically it could happen, but the probability is unknown. At present, the probability of a large-scale increase in cancer or genetic mutations due to ingestion of chlorinated hydrocarbons is unknown. If the probability is less than unity, then the potential effects need to be weighted by the probability of the occurrence.

When probabilities are considered, consequences that are

highly undesirable when they do occur may not figure promi-
nently in the decision if the probability of their occurrence is very
small. The stochastic nature of the variables in the decision
framework makes it extremely important that policy makers take
into account all known probabilities and that they recognize the
probabilities and value implications of actions taken in the
absence of certainty. The criterion adopted in public policy will
almost certainly impose a need to provide solutions that protect
against disaster consequences, no matter how remote. Flood
damages are a case in point. In such a case, rationality may require the absolute value of the consequence to approach infinity,
since the unknown probability of occurrence may approach zero.
Additional ecological research on the impact of pesticides on the
environment is needed to pinpoint the probabilities involved and
to indicate modified actions that reduce the probability and
magnitude of adverse outcomes.

Summary

A decision framework has been developed to show the kinds of
information needed and the relation between the variables involved in pest control policies related to chemical pesticides. It is
possible to conceive of this framework in terms of economic
analysis where all of the values are considered, and to suggest
that social welfare may be maximized without the elimination of
all adverse effects from chemical pesticides.

The areas where knowledge is needed to devise a rational
policy were outlined. These included the effect of pesticides on
positive social benefits, the effect of pesticides on negative social
benefits, a need to develop ways of measuring values outside the
market, and the effects of pesticides on the levels of use of other
resources. Indeed, the latter is important because if pesticides
are in fact a production complement to other resources and not
a substitute, the implications of changes in levels of pesticide use
on the use of other resources can be quite different.

Alternative policy actions and their relation to the framework
were presented. These included changes in the relative price
of pesticides, prohibition of use, financing research to develop

substitute controls and inspection, education, and regulation programs.

Forms of control that provide alternatives to chemicals were examined and the place of each in a pest control program evaluated. The incidence of costs of each alternative was shown to be important in the adoption and use of the various alternatives.

Finally the question of valuation of external effects from pesticide use was discussed and a means of developing surrogate measures of nonmarket effects was suggested. In addition, the role of probabilities in the valuation problem was related to the decision framework.

REFERENCES

1. U.S. Department of Agriculture, Agricultural Research Service, *Research on Controlling Insects Without Conventional Insecticides,* ARS 22–85, October 1963, p. 23.
2. E. F. Knipling, "Alternate Methods in Pest Control," *Symposium on New Developments and Problems in the Use of Pesticides,* Publication 1082 (Washington, D.C.: National Academy of Sciences–National Research Council, 1963), pp. 23–38.
3. U.S. Department of Agriculture, Agricultural Research Service, *Losses in Agriculture: A Preliminary Appraisal for Review,* ARS 20–1, June 1954, pp. vi, 190.
4. E. F. Knipling, *Statement Before the Subcommittee on Reorganization and International Organizations of the Senate Committee on Government Operations,* October 7, 1963, p. 28. Mimeo.
5. R. Van Den Bosch and V. M. Stern, "The Integration of Chemical and Biological Control of Arthropod Pests," *Annual Review of Entomology,* Vol. 7, 1962, pp. 367–86.
6. Stanley A. Hall, "Trends in Insect-Control Agents," *Journal of the Washington Academy of Sciences,* Vol. 52, No. 2, February 1962, pp. 29–35.
7. World Health Organization, *Occupational Health Problems in Agriculture,* Fourth Report of the Joint ILO/WHO Committee on Occupational Health, Technical Report Series No. 246 (Geneva, 1962), pp. 16–25.
8. Wayland J. Hayes, Jr., "Pesticides in Relation to Public Health," *Annual Review of Entomology,* Vol. 5, 1960, pp. 379–404.

3

Conceptual and Measurement Problems in Assessing the Effects of Pesticides Used by Agriculture

At first glance, the evaluation of social benefits from the increased productivity in agriculture resulting from pesticides seems to be a relatively simple task. It appears to involve two steps: first, measure the increased agricultural output resulting from pesticides in physical terms and, second, apply a money value by multiplying the additional quantities of the various products involved by their prices. This, after all, is the common-sensical procedure that an individual farmer would follow in determining benefits from the use of pesticides in his business. The summation of these individual benefits, as a logical continuation of this argument, yields an estimate of the benefits accruing to the community as a whole. In the discussion that follows, the fallacies in this process will be pointed out and the difficulties of correcting these fallacies will be outlined.

One fallacy in this line of reasoning—external costs—has been discussed in Chapters 1 and 2. The error here is that individual decision makers (farmers) may not include all of the costs of pest control in their calculations and, therefore, may overestimate the benefit net of costs. However, even if there were no effects external to agricultural producers—no spillovers into areas of public health or wildlife conservation—this simple approach would yield misleading results under most circumstances. Its deficiencies derive partly from fallacies of composition and partly from the implicit assumption, erroneous except under very special

37

circumstances, that benefits from increased productivity due to the use of pesticides will be manifested wholly in the form of increased output that has value to society. The nature of these inadequacies will become clearer from a discussion of some of the questions that arise in any attempt to use this simple framework for evaluation of gains from pesticide usage.

MONEY RETURNS

One difficulty encountered is the selection of appropriate prices for valuing additional output in money terms. Valuation of extra production presents increasing problems as the size of the unit of analysis increases. At the individual farm level, assuming the conditions of virtually pure competition are fulfilled, which includes a demand for the individual farmer's produce that is invariant with respect to quantity sold, it is correct to value increased output at market prices. Most published estimates of the value of crop or livestock losses due to pests use a similar basis of valuation.

When applied to valuation of aggregate output, however, the use of current market prices (average revenues) to estimate the value of added production due to pesticides can be challenged on two primary points. First, government intervention or other market imperfections and disturbances may have exerted a supporting effect on market prices, causing them to overestimate the social value of increased production. Second, when the problem is one of determining the added revenue accruing to the industry from additional output, marginal revenue and not average revenue (current market price for the present quantity produced) is the relevant value.

Government intervention and the resulting market imperfections were recognized as complicating this issue by Rachel Carson when she drew attention to the cost of surplus-food storage [1]. Market prices do reflect the influence of government price support programs or of monopolistic supply management programs operated by agricultural industries with or without the assistance of government.

It might appear possible to abstract from the effects of government intervention or of a monopolistic market structure by valuing additional production at world prices rather than domestic prices. This basis of valuation has sometimes been used in assessing the effects of agricultural price policies. However, under current conditions of world trade, no easy agreement is possible in many cases on which price, if any, represents the "world price." Indeed, the view that the concept of a "world price" has lost all meaning has formed the basis of far-reaching commodity trade proposals [2]. It is argued that under the present organization of world markets in agriculture, and as a result of the widespread resort to export subsidies and similar trade practices, the levels of prices in world trade are of limited use either as guides to the efficient allocation of resources or as a basis for valuing increments to production.

Even if market prices could be adjusted to abstract from the influence of monopoly elements and trade barriers, their use as a basis for valuing additional output cannot be sanctioned. Earlier we referred to the use of marginal revenues as a basis for valuing increments to production. When an increased quantity of a commodity is offered for sale, total revenue is affected in two ways. First, revenue is increased by the amount received for the "new" output. Second, the price received for all units of the previous production is changed. The second effect derives from the law of demand which calls for a reduction in price (average revenue) as the quantity offered for sale is increased. Therefore, the total revenue resulting from an increase in production is pushed upward by the effects of the larger quantity offered and, at the same time, is pressured downward by the larger quantity being able to command a lower average price. The net effect on total revenue depends upon the percentage drop in the average price relative to the percentage increase in quantity offered (elasticity of demand with respect to prices). In the beginning, we pointed out that individual producers usually face a demand curve that is invariant with respect to quantity sold. However, the aggregate of agricultural producers are believed to face a demand curve where the price varies inversely with quantity. Not only is this demand curve for aggregate agricultural production negatively sloped, but econometric evidence indicates that, historically, a 1 per cent increase in the quantity offered has resulted in a greater than 1 per cent decrease in price, resulting in less total revenue.

On the basis of the previous argument, the errors resulting from the use of average revenue (market prices) to estimate the industry money returns from additional production seem clear. If average revenue is used to evaluate potential increases in production, the money value tends to be overestimated. Conversely, if average revenue is used to value losses, i.e., production saved by pesticides, the return tends to be underestimated. More will be said about this later.

SOCIAL BENEFITS

Let us now turn from money returns to another facet of the problem, namely, social benefits. While the use of marginal revenue as a basis for valuation would be appropriate to estimate the money returns to agricultural producers as a group as the result of additional output, the value might still diverge substantially from social benefits of a pesticide innovation or any other investment or change in policy. There are two main sources of this divergence.

In the first place, in determining social benefits, we are not only interested in the benefits accruing to agricultural producers, but we are also interested in the benefits accruing to consumers of whom agricultural producers are a part. Second, in measuring effects of pesticides on agricultural production, we cannot assume that the effects will be manifested only as increased output. The effects may take the form of savings in resources used to obtain a given level of production. It is reasonable to believe that the effects take both forms simultaneously. Under such circumstances, neither the value of additional output nor the alternative use value (opportunity cost) of resources released separately provides a complete evaluation of the social benefits of pesticides in agriculture, disregarding spillover effects. We need to develop a conceptual framework to include both sets of effects. Moreover, we need to ensure that benefits assessed are those accruing to society, whether they are appropriated through the market or through government price support measures by agricultural producers.

A Conceptual Framework for Assessing Benefits

A scheme for evaluating the social benefits of increased productivity in agriculture has been developed [3] and [4]. If what have been termed the "somewhat unrefined concepts of 'old' welfare economics" are accepted [5], then the demand curve of a commodity can be considered as the curve of marginal value of the commodity to the community and the supply curve can be regarded as denoting the marginal social cost of the resources used to produce the commodity. The area under that part of the demand curve to the left of a given quantity is a measure of total utility of that quantity of the commodity. Similarly, the area under the supply curve to the left of a given quantity indicates the opportunity costs of variable resources used to produce that quantity of the commodity.

Some of the more refined concepts of modern welfare economics [6] can be incorporated into this model, if necessary. Meanwhile, we will abstract from consequential income effects associated with price declines in agricultural commodities as a result of pesticide innovations, and from consequential variations in the prices of factors of production as a result of substitution of pesticides for other inputs and of changes in the relation between input and output. This abstraction will not bend reality seriously in the case of individual agricultural commodities. Income effects and factor-price effects can be serious if we are considering aggregate demand and supply for all agricultural commodities. Accepting the principles or interpretations of supply and demand relationships, the social benefits of an innovation or a change of policy concerning pesticides can be measured by adding, algebraically, the change in total utility or satisfaction (change in the area under the demand curve[s]) for agricultural products to the change in the social cost of resources used in their production (change in the area under the supply curve[s]).

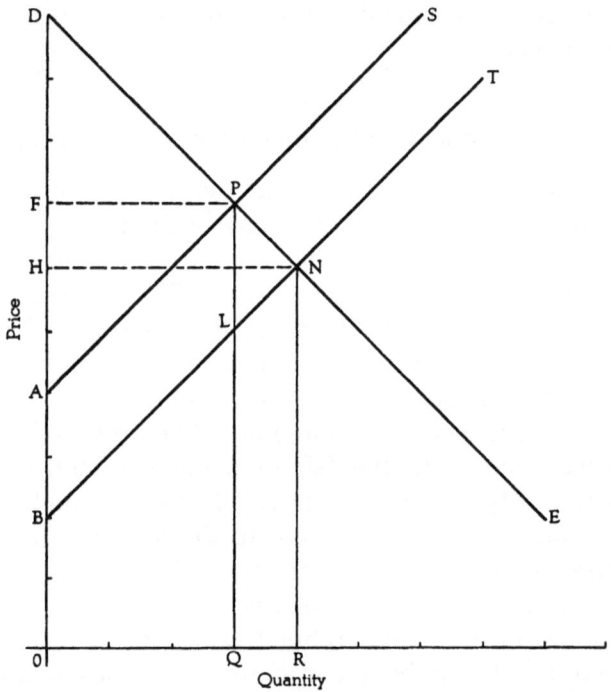

Figure 3. Hypothetical model of demand and supply for an agricultural commodity (competitive market).

A diagram of this model with varying assumptions is presented in Figures 3 and 4. The model in Figure 3 will be examined first. This model presents the price–quantity relationships of a hypothetical agricultural commodity or a small group of commodities. The demand curve is represented by the line *DE* and the supply curve before an innovation or a change in policy concerning pesticides is represented by the line *AS*. We should explain that the analysis does not depend on the linearity of the demand and supply curve nor does it depend on the intersection of the axis by the demand or supply curve. These conditions merely make exposition easier.

Assume that the supply curve shifts to *BT* as a result of the discovery and introduction of a new pesticide. As a result of this shift, demand remaining constant, the quantity produced will increase from *OQ* to *OR* and the price will fall from *OF* to *OH*. Then the area *QPNR*, i.e. (*QLNR* + *LPN*), represents the gain in total utility that was added to the total utility produced by an output of *OQ* defined by the area *ODPQ*. The savings in costs of variable resources is given by the area *BAPL* less the area *QLNR*. Net social benefits resulting from this innovation will be equal to:

$$(QLNR + LPN) + (BAPL - QLNR) = BAPL + LPN.$$

It should be noted that the slopes of the demand and supply curves determine the magnitude of the benefits arising from re-source saving and from increases in satisfaction. As the demand curve *DE* becomes more nearly vertical, that is, the demand for the commodity becomes less elastic, the area of *LPN* becomes smaller. When the demand curve is perfectly vertical, the elasticity of demand being zero, the social benefits are measured entirely in terms of the resources saved. Contrariwise, the more nearly the supply curve approaches the vertical, that is, the supply becomes less elastic with respect to price, the more nearly are social benefits approximated by the increase in total utility or change in the area under the demand curve, *QPNR*.

As shown elsewhere [5], the relationship between crude esti-mates of social benefits obtained by multiplying the actual incre-ment in quantity-produced by the market price referred to earlier, and measurements obtained using the conceptual model outlined above, will vary according to the elasticities of supply and demand. Given the demand and supply elasticities in Figure 3 (not necessarily realistic), the market price method would over-estimate social benefits by an amount equal to the area of *LPN*/2. Clearly, the divergence may be considerable under conditions of inelastic demand and elastic supply, whereas under conditions of infinite demand elasticity (the demand curve being a hori-zontal line) and zero elasticity of supply, the two procedures will yield identical estimates. The condition of infinite demand elas-ticity is more nearly a characteristic of the demand curve facing the farm firm than of the demand for the output of a very large number of firms.

The preceding model and discussion have assumed a competitive market where prices were not supported and where output was not restricted except as determined by supply and demand. In the discussion that follows, a model will be presented incorporating administered prices and controlled output.

In case the products produced are in surplus and, therefore, are subject to some sort of administered price, the diagram in Figure 4 will help in analyzing the effects of pesticides on the measures of satisfaction and resource costs.

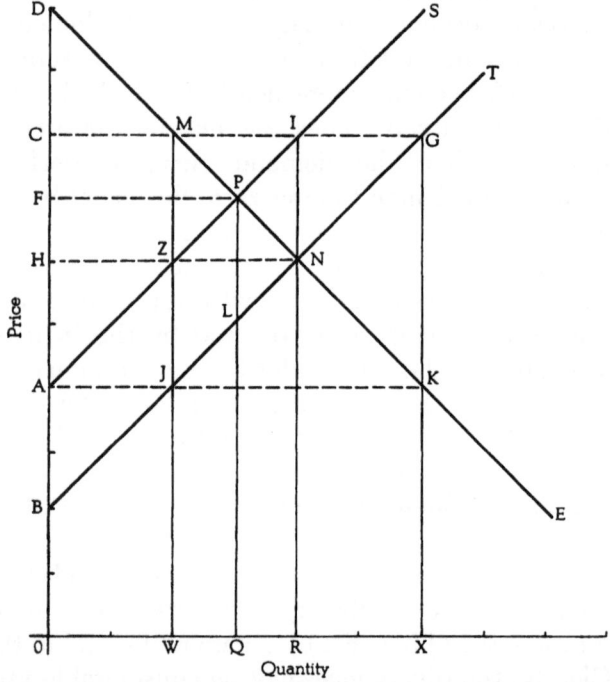

Figure 4. *Hypothetical model of demand and supply for an agricultural commodity (administered prices).*

When the price of the products is administered at the level *OC*, then consumers will take the quantity *OW* and the total satisfaction will be *ODMW*. If the program that administers the price also restricts output to *OW*, then the cost of variable resources committed to production is *OAZW*.

Introduction of a technology such as pesticides shifts the supply curve from *AS* to *BT*, and the cost of variable resources used to produce the commodities represented by *OW* is reduced. Now, the cost of variable resources is represented by *OBJW*. The resources saved as compared to the previous situation are represented by *BAZJ*. The social benefits resulting from the use of pesticides are then equal to *BAZJ*, providing the resources saved can be used elsewhere in the economy.

An alternative type of program produces different results. Again suppose that the price is administered at the level *OC*, but that the output is not restricted to *OW*. Rather, producers are allowed to produce the amount *OR*, since the demand curve is really *DMI* and not *DE*. Such a policy means that the government then takes the amount of output *WR* and places it in storage. The total satisfaction is still represented by *ODMW* and the cost of the variable resources used is represented by *OAIR*. The extra resources used, compared to the previous example, are represented by *WZIR*. As use of pesticides causes the supply curve to shift from *AS* to *BT*, if output is still not restricted, output will expand to *OX* since the demand curve is really *DMIG*. The total satisfaction remains at *ODMW* and the cost of the variable resources used is now increased to *OBGX*, which is larger than *OAIR* by an amount represented by *OBJW*. In this case, society is worse off because more resources are used to produce the same amount of satisfaction, and the extra resources have been withheld from whatever alternative employment might have existed for them.

Under a third type of program, the price to producers is administered at the level *OC* and the price to consumers is determined in the market. In Figure 4, before the introduction of pesticides, this would be represented by the price *OH* for an output *OR* using the supply curve *AS*. The total satisfaction is represented by *ODNR* with the output *OR*. The difference between *OH* and *OC* is a transfer payment to producers. With the introduction of a technology such as pesticides, the supply curve will shift to *BT*. Output now expands to *OX* and the price paid by consumers falls to *OA*. Total satisfaction is now measured by *ODKX*, an increase over *ODNR*. The total cost of variable resources is represented now by *OBGX*, an increase equal to the area of *NIGK*. So, the increased satisfaction is equal to *RNKX* and the increased cost of variable resources is equal to *NIGK*, with the difference being equal to the area of the triangle *KNG*. This difference represents the social benefit, ignoring the cost of the transfer payment and any possible effects the increased farm income might have on other sectors of the economy.

Let us examine this third alternative under some different conditions. Suppose that the introduction of pesticides, as a part of the technology, shifts the supply curve for agriculture from *AS* to *BT* and that, at the same time, the administered price is removed. Output does not expand, but remains at *OR*, and the competitive price *OH* then clears the market. Total satisfaction is still *ODNR*, as before, but the resources used to produce the output *OR* are now valued at *OBNR*, not *OAIR*. The benefit accrues to society in the form of resources saved rather than as increased output. The value of the resources saved is represented by the area *BAIN*. Under these conditions, pesticides are an input-saving rather than an output-increasing technology.

Earlier in this chapter the technique of measuring social benefits by valuing expected output at market prices was criticized. The discussion of the models in Figures 3 and 4 has given logical reasons for errors in social benefits estimated in this manner.

The models presented here do not include costs of research or regulatory activities related to pest control, except as these items are reflected in the cost of pesticide chemicals and equipment for their application. In other words, public research, extension, and regulation expenditures associated with pesticides are not a part of the resource costs considered in the conceptual models. Neither are the spillover effects (external costs and benefits) included in these models, such as those affecting the health, wildlife, or recreation sectors of the economy. Therefore, the social benefits referred to in connection with the models in Figures 3 and 4 are the benefits coming from agriculture, net only of the resource costs. Net social benefits can be obtained by deducting the public research and regulatory expenses from the benefits coming from agriculture and adding or subtracting, as the case may be, any spillover effects outside of agriculture.

Distribution of Benefits

Let us now consider the distribution of the benefits coming from agriculture. Benefits arising from introduction of a technology such as pesticides are not necessarily distributed equally among consumer and producer groups. The total utility (the area under the demand curve to the left of the quantity taken) can be divided into three components. These are (a) the cost of variable resources, (b) the consumer surplus, and (c) the producer surplus. These components can be identified in Figure 3. When the output was OQ, the total utility was represented by the area $ODPQ$. The cost of variable resources was represented by the area $OAPQ$. This leaves the triangle ADP to be distributed between consumers and producers. Consumers turned over to producers an amount of money equal to the area $OFPQ$ which is larger than $OAPQ$ by an amount equal to the triangle AFP. AFP is the producer surplus, which the producers may use to pay for fixed resources that do not change with output such as land, family labor, etc., and is the amount of money received by producers in excess of the cost of variable factors such as fertilizer, hired labor, etc. This leaves the triangle FDP as the consumer surplus, the amount of satisfaction received in excess of what consumers paid for the output OQ, viz. ($ODPQ - OFPQ = FDP$). In Figure 3, the consumers and producers share the surplus equally, that is, $FDP = AFP$. Likewise, when the pesticide technology was introduced, the supply curve shifted down so that BT was parallel to AS. Therefore, the social benefits resulting were equally shared by producers and consumers. However, if the demand curve had been more nearly vertical, given the supply curves, more of the social benefits would have accrued to consumers than to the producers. So the distribution of benefits depends upon the relative supply and demand elasticities.

In Figure 4 the effects of pesticide technology on social benefits were analyzed when the products were in surplus. Let us look at the distribution of benefits under these conditions. The first situation considered output restricted to OW and a price of OC. The total utility was defined as $ODMW$ and the cost of variable resources as $OAZW$. Therefore, the producer's surplus is $ACMZ$ and the consumer surplus is CDM. If pesticides reduce the cost of producing the output OW, then the producer's surplus becomes $BCMJ$ and the consumer's surplus remains at CDM. In other words, all of the benefits accrued to producers because the output was restricted and the cost lowered.

A second situation supported the price at OC, but did not restrict output. In this case the output is OR, yet total utility is equal to $ODMW$. Since output is OR, the cost of variable resources is equal to $OAIR$. Therefore, the producer surplus is equal to ACI and the consumer surplus is equal to CDM. An amount ZMI has been given to producers as a result of the support price program. Under such circumstances, the shift in the supply curve that results from the introduction of pesticides leaves the consumer surplus unchanged and increases the producer's surplus to BCG. Under this alternative, the increases in the producer surplus and payment for the added resources to produce the output in excess of OW come from sources other than the market. In addition, the extra resources used were not available for alternative uses.

A third alternative examined in Figure 4 assumed an administered price to producers equal to OC. Consumers, however, could buy as much of this output as they wished in accordance with their demand curve. Output produced was OR, for which the producers received $OCIR$. Consumers purchased all of the output OR at a price OH. Total utility then was $ODNR$ and the consumer surplus amounted to HDN. The producer surplus in this case was ACI. In this hypothetical situation, $HDN = ACI$. The effects of changes in supply and demand elasticities, discussed earlier, also apply here. Now, the introduction of pesticides, which shifts the supply curve to BT, and the elimination of the support price result in no change in the size of either the

consumer or the producer surplus. Consumers still enjoy the surplus *HDN* and producers receive an amount *BHN* which is equal to *ACI*. The total social benefits from pesticides are represented by resources saved. Whether these benefits are distributed to consumers depends on the use made of the resources saved.

Now it can be seen that, as the benefits from pesticides vary, given different market situations, so does the distribution of the benefits between producers and consumers as market situations and demand and supply elasticities vary.

Applying the Model

As has been observed of the concept of "consumer surplus" [7], our models are clearly of assistance in indicating where we should look for indications and estimates of the social benefits and/or damages. However, is it operational so far as assessing the social benefits from pesticides in agriculture is concerned? It is possible to outline procedures, tried and proven in other areas of agricultural economics, for the estimation of social benefits in accordance with our conceptual framework.

The essential information and the steps in analysis for each commodity or commodity group can be listed as follows:

(a) estimate (or obtain estimates of) the demand function;
(b) estimate (or obtain estimates of) the supply function prior to or without the pesticides whose effects are being evaluated;
(c) estimate (or obtain estimates of) the supply function after adoption of pesticides.

From the shift in the supply curve resulting from the introduction of pesticides, the change in the opportunity cost of resources used in production of the commodity can be determined. The change in total utility is determined from the demand curve and added to the change in social factor cost of production.

The study of demand structure for farm products carried out by Brandow [8] provides a ready-made specification of the principal demand relationships required in our analysis of social gain from the impact of technological progress in pesticides on the agricultural sector. Brandow synthesized a systematic description of relationships between the prices of each of 24 crop and livestock products (or product groups) and the quantities of that product and of all importantly related products. The study has been used as a framework for analyzing the price and demand implications of adjustments in the dairy industry of the Lake states [9]. Its comprehensiveness and description of interdependencies between individual commodities and commodity groups make it invaluable for use in analysis of adjustments, in the evaluation of agricultural price policy alternatives, or in studies of the economics of technological change in agriculture as here envisaged.

Estimation of the supply function and the shift in the supply function resulting from pesticide innovations is more difficult. However, there are several possible methods of doing this. One is to use a function which relates output to input to derive total and marginal (incremental) cost functions and from these to compute the output expected to be supplied as price varies. This forms the basis for a supply function. This procedure can be applied to derive either short-run supply functions (taking all inputs other than the single variable under study—in this case pesticides—as fixed) or long-run supply functions (in which either all or several inputs are variable).

Supply functions obtained in this way are subject to several limitations [10]. Functions relating output to input (production functions) can be used "to provide knowledge of some quantities underlying supply functions" [10]. Although the supply basis relationships so derived give some notion of the possible relation between production conditions in agriculture and supply functions in practice, they are unlikely to correspond to actual response functions. Input–output coefficients and price are, in fact, uncertain because of weather and market movements, and decisions are influenced by goals other than income maximization.

The same weaknesses characterize supply curves synthesized by "price-mapping" studies employing budgeting or linear programming—a group of related techniques that constitute the other main method open for estimating supply functions from production functions or experimental data concerning production functions. The advantages of linear programming for this purpose lie in its flexibility. It can be employed to estimate supply responses to changing product prices including cross-elasticities of supply, response to changes in input prices, and to changes in coefficients (input–output relationships). The effects of a new or improved input, such as pesticides, can be determined, provided one can specify the resulting changes in resource requirements for each activity or process.

If these cannot be specified in advance, it would be possible at least to determine the effects for a range of possible changes in resource requirements. Decision making could thus be assisted by calculating the outcomes resulting from the new or improved input by varying assumptions concerning its impact on resource requirements.

It will be noted, however, that whatever method of estimating supply functions is used, we need knowledge about the technical relationships between inputs of pesticides and output of crops or livestock. These relationships are, in their simplest form, envisaged as a function showing the relationship between quantities of pesticide applied and yields, given the level of other inputs. A concept similar to the production function is the dosage-mortality relationship for pesticides—the percentage kill—achieved in an insect population by varying quantities or concentrations of pesticides applied, being plotted along the vertical axis, instead of crop yield or production.

ESTIMATING PRODUCTION FUNCTIONS

In assessing the effects of pesticides we are, in the simplest sense, merely measuring the difference in production between two (or more) points on a production function resulting from a change in pesticide inputs.

For small technical units, such as an acre of land, the two or more points on a production function may be directly observed from experiments. This is not possible when the unit is very large, e.g., Californian or U.S. agriculture. At this level of aggregation one clearly cannot expect to observe directly points on a production function or surface. When inputs of pesticide chemicals are varied, other inputs will normally be concurrently varied. As a result, direct observations will yield not a true production function, but a contaminated relationship allowing for substitution of mechanical, cultural, and biological control measures. Even if the problems encountered in measuring an agricultural output composed of more than one commodity could be satisfactorily resolved, estimation of the production function by direct observation would generally be impossible. If the differences in yields (due to a complex of such changes) are ascribed simply to changes in the input of pesticides, the relationship obtained consists not of two or more points on the same production function, but rather a combination of movements along, and shifts between, different production functions.

A number of methods might possibly be used to measure the contribution of pesticides to agricultural output. Under some circumstances quasi-experimental observations may be possible. That is to say, yields may perhaps be observed with and without pesticides, with other inputs and conditions unchanged. Thus it may happen that yield differences due to a particular pesticide may be directly observable from the comparison of otherwise homogeneous areas of production that are subject to different pest control practices. This method relies on the fortuitous juxtaposition of homogeneous areas differentiated only by some legal restriction, import or excise duty, or other circumstances affecting the availability of a particular pesticide chemical or the way it can be used. While it may sometimes be valid to impute observed differences in yields of particular commodities to differences in pesticide usage, the method will obviously have, at best, very limited application. It will be rare, indeed, that adventitious differences in pesticide policy between contiguous or physically homogeneous areas are the only, or even the major, source of difference.

Conceivably, the same method could be applied to observations for the same area in successive time periods. That is to say, the short-term effects of a new pesticide could be directly observed at the time of its introduction. In other words, the change in yields taking place following introduction of the pesticides could be observed before anything else had time to change. This method would perhaps be applicable in a fairly static agriculture with administered prices, but under the more dynamic conditions typical of an advanced economy it has patent limitations. In any event, the weather does not take long to change. It may be feasible to eliminate the complicating effects of weather in the data being analyzed by using indexes which have been constructed [11] of the influence of weather on crop yields. However, it should be noted that "weather" as reflected in these indexes covers both direct and indirect effects of weather including "crop damage by certain pests."

Ceteris paribus is, of course, less useful as a description of the real world than as an analytical or expository abstraction. Only in a very limited number of cases will the interpretation of historical data as though they had been generated by a controlled experiment fail to be seriously misleading. In practice, moreover, the comparison of production or yields in like areas subject to unlike influences, so far as a particular input is concerned, involves the risk of imputing unexplained residuals entirely to pesticides. Such radical simplification of the analytical model can only be obtained by placing extremely limiting restraints upon the conditions in which the model can appropriately be employed. Superficially, these conditions may often appear to be fulfilled, while close inspection may show this not to be the case. It has been suggested that this approach might be applied to assessing the effects of pesticides on alfalfa production since nothing else had changed. Actually, development of alfalfa varieties highly resistant to spotted alfalfa aphid and to a nematode affecting alfalfa was found to be one of the outstanding successes achieved in the development of resistant crops [12], to such an extent that attributing all recent gains to chemicals could be grossly misleading.

More sophisticated models seem at first sight to be available in the form of multiple regression-of-time series or cross-sectional data. Functions linear in the logarithms (Cobb-Douglas) have been the usual form employed because of their computing ease, and production functions have been fitted to interfarm, interregional, and international cross-sectional data.

Griliches [13] fitted aggregate production functions (for U.S. agriculture) to 1950 census data of output and input (values) for 68 agricultural regions. Data on quantities or values of pesticide inputs for each of these 68 regions could perhaps simply be plugged into Griliches' models to derive an aggregate production function for U.S. agriculture, with the inputs including pesticides as a separate category.

Nevertheless, one would expect to encounter a number of difficulties in such an analysis. Estimates of production parameters may be seriously disturbed by intercorrelation between inputs. This is especially likely to occur if international data are used to derive an aggregate production function. Pesticide usage, machinery, fertilizers, etc., are probably good joint measures of the level of agricultural technology in various countries. In any case the use of international cross-sectional data calls for even more heroic assumptions concerning the lack of divergence in production functions. It could also add considerably to problems of specification bias, especially as available data on inputs in agriculture are often fairly limited even in some more advanced economies.

While some of the pitfalls encountered in the use of cross-sectional data would be avoided by time-series analysis, other difficulties would replace them. Indeed, the use of time-series data is virtually ruled out of consideration by the problem of shifts in the production function over time. By the time sufficient annual observations have been accumulated to permit statistical estimation of the input–output relationships, the production function itself has substantially changed.

These obvious weaknesses in techniques employing self-generating observations suggest that experimental data might more profitably be used. Unfortunately, formidable aggregation problems are also encountered in this approach. There are a number of well-known sources of potential errors in blowing up experimental data for a technical unit to obtain input–output relationships for a larger unit.

First, experimental observations may relate to a higher standard of performance of associated services than can be expected on farms. Thus, better timing of pesticide applications and more accurate deposition of chemicals may be encountered in experiments than in actual operations.

Second, reported experiments may relate to the benefits of a pest control method under conditions of heavy infestation by a particular insect species or weed. Normal expectations may be for much lower levels of infestation, with the control treatment being undertaken largely as a form of insurance against loss. Thus selection of published experimental data toward the more spectacular results may also lead to serious disturbance of the production function parameters obtained.

Third, experiments often relate to one particular pest as it affects a particular crop. The combined effects of controls for several different pests may not be additive. Indeed, where each pest singly, if not controlled, could virtually eliminate production of a marketable product, simple addition of observed differences could suggest that pesticides were responsible for more than 100 per cent of actual yields. Experimental data needed to establish production surfaces for several inputs are seldom available.

Fourth, the benefits of pesticide applications may often take the form of quality changes rarely subjected to market valuation as part of the experiment. A larger proportion of output may be of higher grades (or simply of better quality in the absence of market grades) or may be marketable at periods of the year when favorable price differentials are received by producers.

Fifth, observations of yields with and without a specific treatment may not measure the full impact of pesticides on production. Farms or plots not treated may still enjoy an umbrella of partial protection provided by control measures on neighboring farms. This is essentially a problem of externalities from one technical unit disturbing, or being confused with, the effects observed for inputs by another technical unit.

A further difficulty arises from the need to include the time factor in the production function for pesticides or other pest control measures. In some experiments, the effects of pesticide applications have been determined by withdrawing treatment from areas previously subjected to control measures for a considerable period. The effect observed in the immediately succeeding production period may either exceed or fall short of the differences observed after a longer period has elapsed. Sometimes this disparity arises because it takes time for natural enemies of the pest to recover from the low levels to which their numbers have been depleted by a succession of insecticidal treatments. At other times the full effects of cessation of pesticide applications may not become immediately apparent because the pest takes time to build up. The problem of cycles or weather-induced fluctuations in the severity of pest infestation and damage may also complicate the experimental assessment of yield effects, as noted above. Rarely will experiments, currently available, have determined a probability distribution for varying degrees of pest damage or for varying yield losses.

Finally, there is one familiar difficulty experienced by agricultural economists in deriving production functions from experimental data for any input. Because the effects of fertilizers or pesticides have not been considered by scientists within the conceptual framework of an economically meaningful production function, experiments will not have been designed to yield observations for several input levels so that the relationship can be delineated over the whole range of likely experience. Typically, experiments seek merely to establish the significance of differences observed at one or a few levels of input only. Yields may be recorded simply "with and without" pesticides, although the single dosage rate will usually be mentioned in the descriptive part of the report. Experiments on toxicity of chemicals generally follow the same pattern, with scientists seeking to establish one or two points on a dosage–mortality relationship, e.g., the LD 50 (lethal dose 50 per cent) or the amount causing mortality to 50 per cent of the exposed population.

The use of experimentally obtained input–output relationships in synthetic approaches to pesticide problems could have, for decision-making purposes, some important advantages over regression analysis. The synthetic (budgeting or programming) approach can be applied to the *ex ante* evaluation of policy alternatives based upon biological data or other technical information concerning the physical or biological effects of chemicals. It does not require prior commercial utilization of a particular chemical before an evaluation of its effects can be made. On the other hand, the statistical method of estimating aggregate production functions, either from time-series or interunit observations, does call for the use of a particular chemical or group of chemicals over a sufficiently long period or on a sufficiently broad scale for statistical evidence of its effect to be reflected in the data.

Short-Cut Methods of Approximation

In most circumstances, as we have seen, the benefits from pesticides (or from any other technological change) will accrue to society both as additional output and as input savings, with the blend dependent upon supply–demand relationships for the commodity. Imputing a value, either to the actual increment in quantity produced or to the actual quantity of resources saved by an innovation, is likely to fail to estimate social benefits accurately. On the other hand, a closer approximation may be obtained simply by estimating the increase in quantity produced, if the quantity of all other inputs remained constant, and valuing added production at some adjusted price. Using a price halfway between the old and new market prices would give a reasonable linear approximation to the increase in total utility (change in area under the demand curve), which exactly corresponds, as we have seen, to the social benefits in the special case where the price elasticity of supply is zero. Alternatively, one could approximate social benefits by calculating the change in the quantity of inputs for the same value of output and valuing this at current or adjusted factor prices. This alternative method gives an estimate of savings in resource costs corresponding to social benefits in the special case where price elasticity of demand is zero.

Ordinarily, the variation between these two extremes may be relatively small. For hybrid corn, Griliches found that a variation in supply elasticity from zero to infinity would result in only a 7 per cent difference in estimates of social benefits and concluded that the elasticities have only a second-order effect and, hence, different reasonable assumptions about them will affect the results very little [14].

Griliches followed the first of the above short-cut methods assuming that hybrid seed increased average corn yields by 15 per cent above those for open-pollinated varieties. Because either hybrid or open-pollinated seed must be used to produce an acre of corn (i.e., the innovation is either used or not used), the evaluation of social benefits in this instance is somewhat simplified. In the case of pesticides we cannot say, quite so meaningfully, that pesticides increase cotton yields by 15 or 30 per cent, unless the quantity of pesticides per acre of cotton, the number of treatments of the crop, and so on, are understood. However, if the quantity and frequency of pesticide applications by farmers are determined by standard extension recommendations or by following the instructions on the label, we could have a situation roughly similar to that for hybrid corn. It would then be unnecessary to know the production function for pesticides (crop or livestock yields for each level of input of the variable resource, pesticides). We would need to know only the average increase in yield obtained by using pesticides in the recommended dose and the area treated. For many crops, actual conditions of pesticide usage would seem to approach this situation. For instance, decisions to carry out insecticidal treatments of major crops in many states are guided, and may even be substantially determined, by releases from natural history survey sections or other economic entomology agencies that give warning to county agents and farmers of conditions favoring insect damage or incipient build-ups in insect populations, and indicate recommended control or preventive programs.

Summary

In this chapter we have reviewed the conceptual and operational problems of assessing the effects of pesticides used by agriculture. The concept of increased output and resources saved by pesticide technology has been introduced and explained. It

was shown that the benefits and the costs of pesticides used in agriculture are dependent upon the responsiveness of output and demand to changes in price.

Several approaches that can be used to estimate the effects of pesticides were presented and their limitations examined. All of these approaches have been used in research concerned with other problems. The tools and the concepts are available to do the assessment job. A shortage of data and the development of surrogate measures seem to be the primary problems. In Chapter 8 some alternative research strategies and possible sources of data are presented, based on the discussion in this chapter.

Evaluation of pesticides as a technology is more complex than the valuation of mechanization or fertilizer. In any event, the methods discussed in this chapter can help to provide crude measures of the impact of pesticides on agricultural output and can furnish information to be used as a basis for research priorities to guide investment in the more detailed research that is necessary to provide answers to specific problems within the total pesticide question.

REFERENCES

1. Rachel Carson, *Silent Spring* (Boston: Houghton Mifflin Co., 1962), p. 19.
2. J. H. Richter, "A Note on the Pisani Plan," *International Journal of Agrarian Affairs*, Vol. 3, No. 5, June 1963, pp. 294–300.
3. Luther G. Tweeten and James S. Plaxico, "Long-Run Outlook for Agricultural Adjustments Based on National Growth," *Journal of Farm Economics*, Vol. 46, No. 1, February 1964, pp. 39–55.
4. T. D. Wallace, "Measures of Social Costs of Agricultural Programs," *Journal of Farm Economics*, Vol. 44, No. 2, May 1962, pp. 580–94.
5. Marc Nerlove, *The Dynamics of Supply* (Baltimore: Johns Hopkins Press, 1958), pp. 222–35.
6. J. R. Hicks, *A Revision of Demand Theory* (New York: Oxford University Press, 1959), pp. vii, 198.
7. Abba P. Lerner, "Consumer's Surplus and Micro-Macro," *Journal of Political Economy*, Vol. 71, No. 1, February 1963, pp. 76–81.
8. G. E. Brandow, *Interrelations Among Demands for Farm Products and Implications for Control of Market Supply*, Pennsylvania State University Bulletin 680, August 1961.
9. University of Minnesota Agricultural Experiment Station, *Equilibrium Analysis of Income Improving Adjustments on Farms in the Lake*

States Dairy Region, 1965, Technical Bulletin 246, October 1963, p. 54.
10. E. O. Heady and J. L. Dillon, *Agricultural Production Functions* (Ames: Iowa State University Press, 1961), pp. 59–64.
11. James L. Stallings, "Weather Indexes," *Journal of Farm Economics,* Vol. 42, No. 1, February 1960, pp. 180–86.
12. E. F. Knipling, *Statement Before the Subcommittee on Reorganization and International Organizations of the Senate Committee on Government Operations,* October 7, 1963, p. 12. Mimeo.
13. Zvi Griliches, "Estimates of the Aggregate Agricultural Production Function from Cross-Sectional Data," *Journal of Farm Economics,* Vol. 45, No. 2, May 1963, pp. 419–28.
14. Zvi Griliches, "Research Costs and Social Returns: Hybrid Corn and Related Innovations," *Journal of Political Economy,* Vol. 46, No. 4, October 1958, pp. 419–31.

4

A General Review of the Agricultural Consequences

This chapter reviews first the technical literature on the nature and incidence of the effects of pesticides on agricultural production. Some typical examples are cited of yield increases and resource savings resulting from the use of pesticides. Next, the functions of pesticides in combating yield instability and some of their general consequences in agriculture are discussed. Finally, since concern has been expressed about the possible pollution of soils, a brief review of the technical literature dealing with decomposition of organic chemicals in soils is presented.

A major part of the experiments conducted to show the effects of pesticides on agricultural production and resource use have been conducted by the U.S. Department of Agriculture. Therefore, it is natural that much of the information that follows is derived from this source. Since the USDA may not approach this task in a disinterested manner, and since there is really very little conclusive evidence from disinterested sources, the estimates of beneficial effects need to be treated critically. While attempts were made to present some evidence of consequences in physical terms, in some cases value estimates of the type criticized in Chapter 3 were the best estimates available.

The authors have screened the references and have attempted to present the most definitive of those available. The review which follows is intended to point to areas where the functions of pesticides seem to be exhibited and to present some idea of the potential magnitude of the effects arising from these functions.

BENEFICIAL EFFECTS OF INSECTICIDES

The benefits of insecticides in agriculture take several forms: increased yields, improvements in the quality of produce, and changes made possible in farm practices and in the location and timing of production. The broad consequences of the latter changes are examined near the end of this chapter. In this section experimental data show the increases in marketable yields of selected crops and livestock products. These examples have been extracted from an extensive compilation of such results by the various agencies of the USDA [1].

Yield Increases

Cotton. Striking differences in cotton yields have been observed over time where insecticide treatment is compared with no treatment. Continuous observations extending over periods of 20 to 37 years are available for selected areas. These are summarized in Table 7.

The comparisons indicated that the use of insecticides since 1945 has been associated with yield increases in these areas averaging from 41 to 54 per cent. The apparent effectiveness of the synthetic organic insecticides is demonstrated by a comparison of results for years before and after 1945. However, it should be remembered that cotton varieties were improved and fertilization

Table 7.   Percentage Increase in Cotton Yields in Insecticide-treated over Untreated Plots, Selected U.S. Locations

			Per cent increase in yields	
Location	Years	All years	Pre-1945	Since 1945
Florence, South Carolina	1928–58	40.6	23.6	53.9
Tallulah, Louisiana	1920–56	31.2	26.4	41.3
Waco, Texas	1939–58	41.8	34.0	53.0

SOURCE: U.S. Senate, *Interagency Coordination in Environmental Hazards (Pesticides),* Hearings Before the Subcommittee on Reorganization and International Organizations of the Committee on Government Operations, United States Senate, Eighty-Eighth Congress, March 1, 1965, Appendix I to Part I, p. 64.

increased during this period. An estimated 66 per cent of the total area planted to cotton was treated with insecticides in 1958. In 1961, it was estimated that about 80 per cent of the cotton acreage in areas where boll weevil damage is prevalent was treated with insecticides at an average cost of $8.10 per acre treated for material and $4.72 for application, making a total of $12.82. This amounts to costs of about $12.00 per bale for chemical insect control [2]. The cost of all chemical control of pests on cotton in the boll weevil areas averaged 5.6 cents per pound of lint cotton marketed in 1961.

Small grains. Sharply increased yields of wheat, oats, and barley have been obtained when parathion has been applied to control heavy infestations of greenbug, an aphid causing widespread damage in these crops. This is illustrated in Table 8 which is compiled from data provided by the USDA.

Sugarcane. Sugarcane grown in Louisiana and Florida is subject to the depredations of the sugarcane borer and about three-

Table 8. Yield Increases in Small Grains Following Greenbug Control with Parathion, Selected U.S. Locations

| Crop | Year | Location | Yields in pounds per acre | | Per cent increase due to treatment with parathion |
			Untreated	Treated	
Wheat	1951	Oklahoma	54	666	1,133
	1952	Oklahoma	762	1,242	63
	1952	Oklahoma	546	1,398	156
	1956	Texas	804	1,254	56
	1956	Texas	1,380	2,134	54
Oats	1951	Oklahoma	976	1,214	24
	1952	Oklahoma	227	1,308	476
	1954	Oklahoma	68	1,361	1,901
Barley	1951	Oklahoma	58	566	876
	1954	Oklahoma	379	1,373	262

SOURCE: R. G. Dahms and E. A. Wood, "Evaluation of Greenbug Damage to Small Grains," *Journal of Economic Entomology*, Vol. 50, 1957, pp. 443–46, and N. E. Daniels *et al.*, "Greenbugs and Some Other Pests of Small Grains," *Texas Agr. Exp. Station Bull. No. 845*, 1956, p. 14.

fourths of the crop is now treated with endrin. Control of this important pest with endrin and other insecticides provides estimated increases of the order of 35 per cent in raw sugar yields per acre [3].

Corn. Several insect pests reduce corn yields substantially unless controlled. One of the most destructive in the corn belt is the European corn borer. Experimental results in Iowa show increases in yields, during 1955, ranging from 13.1 per cent with dieldrin to 27.3 per cent for treatments with endrin [4]. Experiments with various insecticides for the control of rootworms in 1953 indicated gains in yields ranging from 96 to 118 per cent [5].

These experiments measure the difference for which insecticide treatments were responsible under conditions of heavy insect infestation of crops. It is therefore, of course, impossible to generalize directly that insecticides contributed yield increases of this order of magnitude whenever control measures were applied. However they do give some indication of potential reduction in losses suffered by individual producers during serious outbreaks of insect pests.

Other Crops. During 1944 and 1945, control of aphids with DDT on experimental plots of potatoes in Maine increased yields by about 70 per cent [6]. The degree of control afforded by DDT declined steadily, however, in the ensuing 7 years.

Control of alfalfa weevil with heptachlor was associated with increased yields of alfalfa hay in Maryland by from 40 to 92 per cent in 1956 and 1957, depending upon the time of application [7]. In addition, the hay was of much better quality following treatment. The alfalfa weevil is said to be the most important pest of alfalfa in the United States, seriously affecting many areas from east to west.

Use of insecticide granules for control of rootworms, during 1961, is reported to have increased yields of peanuts by 71 per cent [8].

Livestock Products. One of the most serious sources of loss in the cattle industry is the cattle grub. Heavy infestations reduce weight gains and milk production of animals affected. The full magnitude of these losses is difficult to assess. However, information is available on the extent of damage to meat and hides by grubs. Quality discounts and trim losses are estimated at $7.60 per head and the discount of the damaged hides at $3.36 per head, totaling about 4.9 per cent of the carcass and hide value. The Tanners' Hide Bureau estimates that some 22 per cent of the 45 million cattle marketed annually in the United States are affected. The annual loss from this damage to meat and hides is put, on this basis, at $110 million.

Recently, systemic insecticides have provided 90 to 100 per cent control of cattle grubs at a cost of $1 per head. An estimated 6 million cattle were treated with an insecticide in 1962 and this was expected to rise to 10 million in 1963.

Control of horn fly with insecticides provides the only effective protection currently available and results in substantial gains throughout the United States. In Kansas, DDT sprays and dips resulted in gains of from 30 to 70 pounds per animal [9]. The use of DDT for horn fly control is no longer recommended and other insecticides such as toxaphene, methoxychlor, and malathion are now widely used for this purpose.

During the years 1955 to 1957, butterfat production of Illinois dairy cows treated for control of stable flies was found to be from 6.5 to 29.8 per cent higher than that of untreated cows [10].

Changes in Quality

Insecticides also have important effects in improving the quality of agricultural products. Quality effects are more difficult to assess than the physical increases in yields. However, an indication of their importance in fruit and vegetable crops can be obtained from published experimental data. In many cases the improvement in quality makes the difference between a marketable crop and the need to destroy all or most of the fruit harvested.

Where no control measures are taken, codling moth damage to apples and pears can range from 50 per cent to near total loss of the crop. Modern insecticides, especially guthion and Sevin, give good control and usually reduce the worm count per hundred

Table 9. Examples of Increased Crop Yields from Chemical Weed Control, Selected U.S. Locations

Crop	Location	Weed problem	Herbicides used	Without herbicides	With herbicides
Corn	Mississippi	annual weeds	atrazine + 2,4-D	90	100 bu
	Iowa and Illinois	annual weeds	atrazine + 2,4-D	100	108 bu
Soybeans	Missouri, Iowa, and Illinois	annual weeds	amiben	36	40 bu
	Mississippi	annual weeds	prometryne	30	37 bu
Rice	Arkansas	annual weeds	propanil	2,694	4,683 lb
	Mississippi	annual weeds	propanil	2,929	4,794 lb
	California	annual weeds	propanil	4,565	6,211 lb
Wheat	Ohio	annual weeds	2,4-D	32	36 bu
	Montana	perennial weeds	2,4-D	47	65 bu
Sugar beet	Minnesota	annual weeds	DATC	13	14 tons

SOURCE: U.S. Senate, *Interagency Coordination in Environmental Hazards (Pesticides)*, Hearings Before the Subcommittee on Reorganization and International Organizations of the Committee on Government Operations, United States Senate, Eighty-Eighth Congress, March 1, 1965, Appendix I to Part I, p. 83.

apples to less than one. The proportion of wormy fruit in California in 1956 was 21 to 23 per cent without insecticides and 0.5 per cent using guthion. Damage to untreated fruit is very much higher in many areas.

A reduction of the same order of magnitude in damage to peaches by Oriental fruit moth results from the use of parathion or guthion. Damage resulting in unmarketable crops of sweet corn also results from failure to control corn earworm. Control with insecticides in Texas, Florida, Illinois, Arkansas, and Missouri has resulted in increases ranging from 58 to 91 worm-free ears per hundred. The development of these control methods had made possible the rapid development of sweet corn production in Florida and the Rio Grande Valley, as well as increasing yields in other United States areas.

BENEFICIAL EFFECTS OF HERBICIDES

Herbicides in agriculture increase yields by reducing competition between crops and weeds for water, light, and plant nutrients and by controlling weeds that harbor insect pests and fungus diseases or are toxic to livestock. They also save resources, including the labor, machinery services, and other inputs involved in mechanical weed control and in the cleaning of grains and seeds in processing or in commercial seed houses.

Some examples of increased crop yields due to control of weeds by herbicides are set forth in Table 9, compiled by the USDA [1].

Table 10 presents similar experimental data for observed yield increases from the use of herbicides in vegetable production, while in Tables 11 and 12 the USDA has summarized gains in yields of rangelands and permanent pastures respectively.

Input Savings

It is estimated that chemical weed control practices reduced labor requirements on about 1 million acres of cotton grown annually in Mississippi by some 20 hours per acre. As a result, weed control costs were reduced by an estimated $10 per acre

Table 10.   Examples of Increased Yields in Vegetable Production Resulting
from Use of Herbicides

Crop	Weeds involved	Herbicide used	Per cent increase in yields in experimental plots
Carrots	annual weeds	amiben	157
Lettuce		amiben	86
Onions	annual weeds	CIPC	561
Spinach	pigweed, purslane, henbit, crabgrass, barnyardgrass	CIPC	12
Sweet potatoes	annual weeds	CDAA	27
Tomatoes	pigweed, lambsquarter, barnyardgrass, foxtail, plantain	Solan	140

SOURCE: U.S. Senate, *Interagency Coordination in Environmental Hazards
(Pesticides),* Hearings Before the Subcommittee on Reorganization and Interna-
tional Organizations of the Committee on Government Operations, United States
Senate, Eighty-Eighth Congress, March 1, 1965, Appendix I to Part I, p. 84.

Table 11.   Benefits from Chemical Control of Brush on Rangelands

Crop	Weeds involved	Herbicide	Yield per acre (lb) Without herbicide	With herbicide	Per cent increase
Forage, airdry	shinnery oak	2,4,5-T	635	1,640	258
Forage, drymatter eaten by cattle	blackjack oak and associated species	2,4,5-T	2,200	7,100	323
Forage, overdry	low-grade hard-woods (Ozarks)	2,4,5-T	220	1,210	550
Grass, airdry	sagebrush	2,4-D	526	2,075	394
Herbage production	sagebrush	2,4-D	159	502	316
Grass, average production	mesquite (Arizona)	2,4,5-T	157	645	411
Forage grasses	port-blackjack oak	2,4,5-T	223	1,290	578
Animal unit months	sagebrush	2,4-D	702	1,372	195
Forage, airdry	mixed hardwoods	2,4,5-T	164	1,796	1,095

SOURCE: U.S. Senate, *Interagency Coordination in Environmental Hazards (Pesti-
cides),* Hearings Before the Subcommittee on Reorganization and International
Organizations of the Committee on Government Operations, United States Senate,
Eighty-Eighth Congress, March 1, 1965, Appendix I to Part I, p. 85.

Table 12. Benefits from the Control of Weeds on Open Permanent
Pastures

Crop	Weeds involved	Herbicide	Yield per acre (lb) Without herbicide	With herbicide	Per cent increase
Forage herbage	perennial and annual	2,4-D	1,100	2,800	254
Ladino clove	curly dock	4-(2,4-DB)	2,800	6,000	214
Forage grass	perennial and annual	2,4-D	2,000	4,000	200
Forage	undesirable grasses	TCA	2,100	5,400	257
Alfalfa	winter annuals	CIPC	3,000	4,600	153
Birdsfoot trefoil	annuals	dalapon	80	3,860	4,825
Alfalfa	annuals	4-(2,4-DB)	460	1,750	380

SOURCE: U.S. Senate, *Interagency Coordination in Environmental Hazards (Pesticides)*, Hearings Before the Subcommittee on Reorganization and International Organizations of the Committee on Government Operations, United States Senate, Eighty-Eighth Congress, March 1, 1965, Appendix I to Part I, p. 85.

and, in addition, mechanical harvesting operations were facilitated and the quality of the lint improved [11], [12].

Labor savings of 14 hours per acre and a net reduction of $9 per acre in costs of production of peanuts were recorded in Georgia [13]. In sugar beet production in Minnesota, the use of trichloroacetic acid to control annual weeds effected a cost reduction of $10 per acre [14].

Herbicides also permit major savings in the maintenance of irrigation and drainage canals. A survey of 47 irrigation districts and irrigation companies in 17 western states was carried out in 1958 [15]. An estimated net saving of $7.75 million was achieved by the use of herbicides (including 2,4-D and aromatic solvents) and other methods of weed control. Drying, handcutting, clearing, chaining, draglining, and chemical treatment were employed to control aquatic weeds and handcutting, mowing, burning, and spraying for ditchbank weeds. In this computation, water losses prevented were valued at actual costs to water users per acre-foot and allowances were also made for damage to crops, canals,

Table 13. Examples of Benefits from Use of Fungicides to Control Plant Diseases during Production

Crop	Disease	Fungicide	Yield		Quality		Yield increase per cent
			Untreated	Treated	Untreated	Treated	
Apple	scab	captan	2 ton/acre	8 ton/acre	1 ton/acre, good	7.0 ton/acre, good-fancy	300
Apple	powdery mildew	Karaltane	3 ton/acre	6.5 ton/acre	2 ton/acre, good	6.0 ton/acre, good-fancy	117
Peach	brown rot	captan	3.5 ton/acre	6.8 ton/acre	1.7 ton/acre, good-fancy	6.5 ton/acre, good-fancy	126
Peach	leaf curl	ferbam	1 ton/acre	7 ton/acre	1 ton/acre, poor-cull	6 ton/acre, good-fancy	600
Peach	scab	captan	0.5 ton/acre	7 ton/acre	0.5 ton/acre, poor-cull	6.5 ton/acre, good-fancy	1,300
Cherry	leaf spot	dodine	1,700 lb/acre	2,600 lb/acre	1,300 lb/acre, good-fancy	2,600 lb/acre, good-fancy	53
Grape	black rot	ferbam	1,000 lb/acre	8,000 lb/acre	—	—	700
Pecan	scab	dodine	100 lb/acre	300 lb/acre	50 lb/acre, poor-good	150 lb/acre, fancy	200

Tomatoes	early blight and gray leafspot	maneb	4.6 ton/acre	18.1 ton/acre		—	293
Cucumber	scab	maneb	51 bu/acre	148 bu/acre	31% U.S. 1	74% U.S. 1	190
Potatoes	late blight	maneb	360 cwt/acre	418 cwt/acre		—	16
Potatoes	late blight and early blight	maneb	258 cwt/acre	402 cwt/acre		—	56
Potatoes	verticillium wilt	vafam	74 cwt/acre	330 cwt/acre		—	347
Potatoes	scab and scurf	PCNB	232 cwt/acre	300 cwt/acre	40 cwt/acre, U.S. 1	224 cwt/acre, U.S. 1	29
Sweet corn	helminthosporium blight	zineb	862 lb/acre	1,724 lb/acre	385 lb/acre, fancy	1,339 lb/acre, fancy	100
Lima beans	downy mildew	maneb	2,180 lb/acre	3,855 lb/acre		—	77
	anthacnore	zineb	82 bu/acre	355 bu/acre		—	333
Onions	downy mildew	zineb	125 lb per 300 bulbs	217 lb per 300 bulbs		—	70
Corn	seedling blight	thiram	87.1 bu/acre	95.9 bu/acre		—	10
Cotton	seedling diseases	caresa	2,487 lb/acre	2,750 lb/acre		—	11

SOURCE: U.S. Senate, *Interagency Coordination in Environmental Hazards (Pesticides)*, Hearings Before the Subcommittee on Reorganization and International Organizations of the Committee on Government Operations, United States Senate, Eighty-Eighth Congress, March 1, 1965, Appendix I to Part I, p. 92.

farmland, and structures resulting from weeds in irrigation systems.

A Bureau of Reclamation report [16] estimated that in one region costs of controlling annual and perennial weeds on irrigation canal banks was reduced from $35.74 to $22.15 per mile by the use of the herbicides 2,4-D and 2,4,5-T. Use of aromatic solvents to control submersed aquatic weeds in irrigation canals and laterals reduced costs by $330.31 to $38.11 per mile.

BENEFICIAL EFFECTS OF OTHER PESTICIDES

Other chemicals used in agriculture include fungicides to control plant diseases, nematocides, defoliants, and growth regulators. More than 7 million acres of cotton were treated with harvest-aid chemicals in 1962. The combined benefits of mechanical harvesters and defoliants ranged from $21 per acre for spindle pickers to $27 to $33 for stripper pickers [17]. A large part of the decrease in harvesting costs is attributable to the use of chemicals.

Table 13, prepared by the USDA, gives examples of estimated yield and quality improvements resulting from the use of fungicides. Table 14 provides similar data for yield improvements from chemical control of nematodes.

Growth regulators are widely used on tree fruits to thin fruit settings and to prevent premature drop. Estimates of benefits to individual producers range from $10 per acre for drop prevention of Italian-type prunes to as much as $200 per acre from reduced thinning costs and increased output of marketable fruit, in the case of apples.

PESTICIDES AND THE ORGANIZATION OF AGRICULTURE

Pesticides are an integral part of far-reaching technological changes leading to major adjustments in the economic organization of agriculture. More effective control of certain key insect pests has enabled the development of specialized fruit and vegetable production in areas with interregional advantages. This has profoundly affected the efficiency of resource use, the

Table 14. Typical Examples of Crop Yield Benefits and Improved Quality Derived from Use of Nematocides to Control Nematodes

Crop	Nematode involved	Nematocide used	Yield Without	Yield With	Per cent increase
Carrot	root-knot	ethylene dibromide	20 tons	26 tons	33
Celery	root-knot	ethylene dibromide	18.6 tons	28.5 tons	53
Corn	stubby root, stylet and other	ethylene dibromide	25.2 bu	75.1 bu	198
Corn	stubby root, stylet and other	dichloropropene and dichloropropane	25.2 bu	67.0 bu	166
Cotton	root-knot	dibromochloropropane	514 lb lint	1,041 lb lint	100
Cotton	sting nematode	dibromochloropropane	203 lb seed cotton	1,566 lb seed cotton	670
Cotton	sting nematode	ethylene dibromide	203 lb seed cotton	1,605 lb seed cotton	690
Cucumber	root-knot	dibromochloropropane	4.81 tons	11.98 tons	149
Okra	root-knot	dibromochloropropane	2.35 tons	8.00 tons	240
Peanuts	sting nematode	EN 18133 (Cyrem)	787 lb	2,816 lb.	258
Tobacco	root-knot	D-D	1,380 lb	1,672 lb	21
Tomato	root-knot	ethylene dibromide	3,040 lb	10,350 lb	240
Tomato	root-knot	dichloropropane and dichloropropene	6.38 tons	14.26 tons	123
Strawberry	root-knot	ethylene dibromide	137 crates	211 crates	54
Sugar beet	sugar beet nematode	dichloropropene and dichloropropane	10.76 tons	28.10 tons	159
Sugar beet	root-knot	dichloropropanes	11.13 tons	27.48 tons	141

Source: U.S. Senate, Interagency Coordination in Environmental Hazards (Pesticides), Hearings Before the Subcommittee on Reorganization and International Organizations of the Committee on Government Operations, United States Senate, Eighty-Eighth Congress, March 1, 1965, Appendix I to Part I, p. 97.

geographical distribution of production, and the organization of farming and even of the marketing process.

Unprecedentedly rapid changes in agricultural technology and farm organization during the last few decades have led to the coining and widespread currency of the term "technological revolution in American agriculture." It would be overstating the impact of the newer pesticides to claim that they alone have set these changes in motion. At the same time, however, it would be understating their importance to pretend that the pace and form of adjustment could have been maintained without them.

It is unnecessary, for our purposes, to speculate about the question of whether advancing technology in pesticides initiated these changes or was itself induced by technical progress in mechanization and other fields. Whether modern herbicides were developed to order in answer to the needs of a changing agriculture or whether they created a market for machinery of larger capacity is largely immaterial to the pesticide problem outlined earlier.

The role played by pesticides in the technological revolution in agriculture may be clarified by considering three related aspects: the "insurance function," the substitution of pesticides for other inputs, and the contribution of pesticides to interregional shifts in agricultural production.

Pesticides are used by farmers not only to increase yields, but also to stop losses. Insecticides and herbicides serve to reduce greatly the degree of production instability to which the individual producer is exposed. They are often regarded by farmers as a form of insurance against crop damage. By reducing the variance of yields, pesticides may exert a major influence upon the enterprise combinations on farms. The margin of protection they afford may, in many circumstances, avoid the need for hedging against yield uncertainty by diversification of enterprises. Reduction of risk premiums or of discounting for uncertainty operates towards a more efficient allocation of resources. Internal and external limitations on capital investment by the farm firm will tend to be reduced the more assured yields become. Efficiency may consequently be increased by economies of both proportionality and scale.

Farmers are also motivated to use pesticides by the desire to save or economize on other resources, including labor and land. Herbicides substitute for cultivation and their increasing usage often reflects growing difficulties in securing an adequate supply of casual labor for cultivation and weeding, as well as rising costs for labor. Technical considerations, such as the advantages in continuous corn operations and of avoiding excessive compaction of the soil by overfrequent cultivations, also encourage factor substitution of this kind.

Larger-scale cropping operations that become economic by reason of technical developments in farm machinery, for example, may add to the uncertainties associated with reliance upon casual labor and consequently tend to encourage substitution of chemical methods of weed control. Substitution of pesticides for other inputs may be induced either by changing relative factor prices or by changing input substitution ratios (labor-capital substitution) resulting from technological change. Complementarity between herbicides and machinery inputs is distinctly possible in such circumstances.

The two influences just discussed—modification of production uncertainties (including risks uninsurable because of lack of suitable facilities for comprehensive crop insurance) and the advantages of substituting herbicides, defoliants, and other pesticides for other inputs (associated with changing relative factor prices and technological change)—have together contributed to changes in the regional distribution of production of several major crops. Such shifts include the westward movement in the cotton industry.

PESTICIDE RESIDUES IN INTERNATIONAL TRADE

One problem that has already arisen and that may well become more serious with the increasing use of pesticides around the world is the effect of toxic residues on foodstuffs upon their freedom of movement in international trade. Dormal and Hurtig [18] observe that tolerances established in Europe are, for many chemical compounds, lower than in the United States. "These growing difficulties," they state, "are being regarded with increasing apprehension as a hindrance to the free international movement of food and are causing fear in some quarters that, unscrupulously employed, they may be used as trade barriers to restrict trade."

Some form of international standardization of tolerances for toxic residues would seem to offer a solution to this problem. However, when the differing conditions and requirements for pest control in various parts of the world are considered, achievement seems impracticable even in a broad sense. Climatic conditions, the nature and habits of pests, and the varying degrees of population pressure on food supplies are all likely to impose substantial differences. Furthermore, if residue tolerances are based on permissible intakes of toxic substances, differences in the composition of diets among countries would lead to variations in the tolerances derived. A higher residue on, say, rice could be more safely permitted in the United States than in Monsoon, Asia, because it represents a lower proportion of total foodstuffs consumed in the United States.

These difficulties were recognized by the joint meeting of the Food and Agriculture Organization and the World Health Organization experts in 1961, who, however, felt that although different tolerances for any one food may be established in different countries this "does not necessarily impede the free movement of that food in international trade" [19].

Tolerances are maximum permitted levels of concentration of pesticide chemicals or their derivatives on foods, and normally most of a commodity produced will be well inside these limits. Substantial proportions of output could thus move into international trade unless zero or prohibitively low tolerances for chemicals widely used locally applied in importing countries. While possible, this is perhaps unlikely to be a frequent situation if common principles are used by various countries for the determination of residue tolerances.

The need for standard procedures of residue identification and measurement and for international exchange of information on developments in food-testing techniques has also been noted in the specialists' report of the Food and Agriculture Organization and World Health Organization.

PESTICIDE RESIDUES IN THE SOIL

The increasing quantities of chemicals applied directly to the soil to control weeds and nematodes and the deposition of large quantities of insecticides and herbicides applied in above-ground pest control practices have caused some concern as to the possible effects on populations of beneficial soil organisms and chemical properties of soils.

A detailed review of the influence of pesticide residues on the microbiological and chemical properties of soils, with a bibliography of some 207 titles, has been prepared [20]. This literature indicates that many organic pesticide residues are utilized as a source of carbon and energy by certain soil organisms. However, many of the chlorinated hydrocarbon group of chemicals, especially DDT, are highly resistant to decomposition by microbial organisms and, it is believed, the loss of aldrin, dieldrin, heptachlor, and similar compounds from the soil may take place principally by way of volatilization.

It was found in 1950 that one-half of the amount of benzene hexachloride (BHC) applied to the soil was still present after 3 years. This chemical (and, to a lesser extent, lindane) has caused problems of off-flavor in potatoes and canned carrots, tomatoes, lima beans, peaches, and plums [21]. Peanuts are capable of accumulating considerable amounts of BHC from the soil. There are several recorded instances where sweet corn had to be destroyed when grown on land where alfalfa stands were treated the previous year with a half-pound of lindane per acre. These occurrences focus attention on the potential hazards of preferential species absorption of chemicals from the soil and the need for care in using BHC and lindane on nonfood crops grown in rotation with food crops. Recent studies with radioactive tracers have found evidence of uptake of dieldrin and other chlorinated hydocarbons by field crops in the stems and foilage. Therefore, a certain lack of concern with this aspect in the past may have been related to chemical extraction and identification techniques.

Nevertheless, it is the view of some that the vast majority of organic pesticides or their degradants will be found to be subject to microbial decomposition. It should be noted, however, that there has been far more attention, in research, to the decomposition of herbicides in the soil than to processes affecting other kinds of pesticides. There is, accordingly, a need for more research as part of a broader program, on the manner and pace of decomposition and volatilization of pesticides in soil.

Some pesticides, especially soil fumigants and fungicides, initially kill large numbers of the soil organisms, but re-establishment is quickly achieved by organisms surviving the treatment, or from nontreated areas. If the trend toward more widespread use of these materials continues, the dependence on nontreated areas for organism build-up may be questioned. Some organisms reach very high numbers under the conditions of reduced competition following treatment and by utilizing the dead bodies of organisms killed as well as the pesticide residues. Later other organisms develop, and there is a gradual return to a normal composition of the microbial population. This is the sequence when a kill of microbes occurs. However, most of the common insecticides and herbicides do not kill or reduce the activity of soil organisms.

It has also been observed that some pesticides release soluble plant nutrients from the soil both by decomposition of the chemical (and of any initial kill of microbes) and by stimulation of microbial activity in soil organic matter. Approved pest control practices are usually favorable to plant growth, but may retard it if insufficient time is allowed for highly phytotoxic residues to decompose or volatilize, or if a parasitic microbe is the first to recolonize in a fumigated soil, or if, as is possible in some soils, soluble manganese or some other trace element is increased to the point of becoming toxic.

Some scientists consider that any such effects encountered are generally temporary and readily corrected and that the world's soils need not and will not be made sterile or rendered permanently infertile by residues from modern insecticides. However, in 1964 the USDA began a program of monitoring for pesticide residues in soil and water. This program has resulted in a network of locations across the country allowing historical pesticide applications to be used in the study of the effects of pesticides on health.

REFERENCES

1. U.S. Senate, *Interagency Coordination in Environmental Hazards (Pesticides)*, Hearings Before the Subcommittee on Reorganization and International Organizations of the Committee on Government Operations, United States Senate, Eighty-Eighth Congress, March 1, 1965, Appendix I to Part I.
2. E. L. Langsford, *Extent and Cost of Using Chemicals in Cotton Production*, U.S. Department of Agriculture, Economic Research Service, ERS–155, March 1964, p. 7.
3. S. D. Hensley *et al.*, "Field Tests with New Insecticides for Control of Sugar Cane Borer in Louisiana in 1959," *Journal of Economic Entomology*, Vol. 54, No. 6, pp. 1153–54.
4. H. C. Cox, W. G. Lovely, and T. A. Brindley, "Control of the European Corn Borer with Granulated Insecticides in 1955," *Journal of Economic Entomology*, Vol. 49, No. 6, pp. 834–38.
5. J. H. Lilly, "Insecticidal Control of the Corn Rootworm in 1953," *Journal of Economic Entomology*, Vol. 47, No. 4, pp. 651–57.
6. T. E. Bronson, Floyd F. Smith, and G. W. Simpson, "Control of Aphids on Potatoes in Northeastern Maine," *Journal of Economic Entomology*, Vol. 39, No. 2, pp. 189–94.
7. B. A. App, "Studies on Control of Alfalfa Weevil Larvae in the East," *Journal of Economic Entomology*, Vol. 52, No. 4, pp. 663–66.
8. G. M. Boush, M. W. Alexander, and W. L. Powell, "Field Tests With New Insecticides for Control of the Southern Corn Rootworm Attacking Peanuts in Virginia," *Journal of Economic Entomology*, Vol. 56, No. 1, pp. 15–18.
9. E. W. Leake, "DDT for the Control of the Horn Fly in Kansas," *Journal of Economic Entomology*, Vol. 39, No. 1, pp. 65–68.
10. W. P. Bruce and G. C. Decker, "The Relationship of Stable Fly Abundance to Milk Production in Dairy Cattle," *Journal of Economic Entomology*, Vol. 51, No. 3, pp. 269–74.
11. V. C. Harris, "Weed Control in Cotton Over a Ten Year Period by Use of the More Promising Materials and Techniques," *Weeds*, Vol. 8, 1960, pp. 616–24.
12. G. T. Holstein, Jr. *et al.*, "Weed Control Practices, Labor Requirements and Costs in Cotton Production," *Weeds*, Vol. 8, 1960, pp. 232–42.
13. E. W. Hauser *et al.*, "Herbicides and Herbicide Mixtures for Weed Control in Peanuts," *Weeds*, Vol. 10, 1962, pp. 139–44.
14. R. N. Andersen, *Annual Report, Weed Investigations—Agronomic Crops*, Crop Protection Research Branch, Crop Research Division, Agricultural Research Service, U.S. Department of Agriculture (St. Paul, Minnesota, 1962).
15. F. L. Timmons, *Weed Control In Western Irrigation and Drainage Systems*, Joint Report, U.S. Department of Agriculture Agricultural Research Service and U.S. Department of the Interior Bureau of Reclamation, ARS 34–14, September 1960, p. 22.
16. U.S. Department of the Interior, Bureau of Reclamation, *Annual Regional Weed Control Report for 1962* (Salt Lake City, Utah).

17. G. B. Crowe and H. R. Carns, *The Economics of Cotton Defoliation,* Mississippi Agricultural Experiment Station Bulletin No. 552 (1957).
18. S. Dormal and H. Hurtig, "Principles for the Establishment of Pesticide Residue Tolerances," *Residue Reviews,* Vol. 1, 1962, pp. 140–51.
19. World Health Organization, *Principles Governing Consumer Safety in Relation to Pesticide Residues,* Technical Report Series No. 240, (Geneva, 1962).
20. James P. Martin, "Influence of Pesticide Residues on Soil Microbiological and Chemical Properties," *Residue Reviews,* Vol. 4, 1963, pp. 96–129.
21. C. H. Mahoney, "Flavor and Quality Changes in Fruits and Vegetables in the United States Caused by Application of Pesticide Chemicals," *Residue Reviews,* Vol. 1, 1962, pp. 11–23.

5

Effects on Human Health

Actual or potential effects of pesticides on human health take many forms. Better control of insects, weeds, and other pests in food production and storage may bring about improved nutrition by increasing the quantity, quality, variety, and seasonal availability of foodstuffs. However, it would be double-counting to include the effects on agricultural output in the calculus both as a factor contributing to health and as an additional contribution to welfare over and above those effects on total utility reflected in the demand functions for food and fiber.

Pesticides also have important implications for health through their impact on mortality and morbidity rates from insect-borne diseases and, to a lesser extent, through control of allergy-inducing weeds. On the other hand, pesticides are often toxic to humans. Harmful effects on health may arise from occupational hazards in the formulation and application of pesticides; from accidental or incidental exposure of members of the public to toxic substances; and from toxic residues on food or raw materials following treatment of plants, livestock, and food containers, or environmental pollution by pesticides.

In this chapter the evidence concerning the nature of these various effects on human health is examined. The institutional arrangements for protecting human populations from harmful consequences of pesticides (i.e., of reducing the probabilities of harmful effects) are reviewed and evaluated. The next chapter presents a brief examination of conceptual and measurement problems encountered in assessing the social benefits and costs of health programs, together with a review of some of the literature of public health economics, especially that concerned with valuation of changes in mortality and morbidity of human populations.

EFFECTS ON DISEASES

Insecticides have contributed to the control of an impressive list of diseases including malaria, filariasis, dengue, urban yellow fever, virus encephalitis, louse-borne typhus, bacillary and amoebic dysentery and diarrhea, leishmaniasis, bartenellosis, onchocerciasis, sandfly fever, trypansomiasis, yaws, infectious conjunctivitis, cholera, Chagas' disease, scrub typhus, scabies, rickettsialpox, tick-borne relapsing fever, Rocky Mountain spotted fever, and tularemia [1].

Insect-borne diseases of serious incidence in the United States are malaria, Rocky Mountain spotted fever, and encephalitis. Others less common, but potentially of importance, include plague, tularemia, murine typhus, and yellow fever. Estimates of the number of deaths caused by these diseases in the United States are given in Table 15. Reported cases of some notifiable insect-borne diseases are set out in Table 16.

Table 15. Deaths from Some Insect-Borne Diseases, United States

Year	Malaria	Rocky Mountain spotted fever	Tularemia	Arthropod-borne encephalitis
1940	1442	83	189	n.s.a.[a]
1945	443	128	122	n.s.a.[a]
1950	76	31	15	65[b]
1955	18	8	9	32
1960	7	11	4	n.s.a.[a]

[a] Not separately available.
[b] Data for 1952.
SOURCE: U.S. Department of Health, Education and Welfare, Public Health Service, Communicable Disease Center, *Morbidity and Mortality Weekly Report,* various issues.

Table 16. Reported Cases of Selected Notifiable Diseases, United States

Year	Dengue	Malaria	Rocky Mountain spotted fever	Tularemia	Murine typhus
1930	203	98,491	197	661	511
1935	582	137,513	492	782	1,287
1940	66	78,129	457	1,620	1,878
1945	106	62,763	472	900	5,193
1950	26	2,184	464	927	685
1955		522	295	584	135
1960		72	204	390	68
1961		73	219	365	46

SOURCE: U.S. Department of Health, Education and Welfare, Public Health Service, Communicable Disease Center, *Morbidity and Mortality Weekly Report*, various issues.

Notable declines in the incidence of insect-borne diseases have taken place since the introduction of synthetic organic pesticides. For two of the mosquito-borne diseases—malaria and dengue— the downward trend in mortality and morbidity antedates the introducton of DDT. DDT hastened the downward trend of malaria incidence and, superimposed on other factors occasioning the decline, led to its virtual disappearance [2]. This was typical of many countries. However, despite the relatively low number of deaths due to malaria in the United States, malaria deeply affected the life and economic welfare of some 27 million people in 13 southern states, where the disease was endemic. One estimate puts the annual loss in the southern states at $500 million. The basis or source of this figure is not given but almost certainly originated with Williams [3].

A description is available of the methods used in the antimalaria campaign since 1945, when DDT was first used for this purpose in the southeastern states. Spraying internal walls, ceilings, and eaves of houses with an aqueous emulsion of DDT was the principal measure undertaken. This spraying was originally carried out twice a year but, in most states, the schedule was subsequently changed to a single annual treatment. Antilarval or drainage operations were carried out only where necessary to protect communities of 2,500 inhabitants or more. Measures to "build malaria out" by eliminating breeding sites, created by human action, were also enforced.

Details of materials and labor used in the first 5 years of the campaign and of the number of houses treated are set out in Table 17. The figures for reported cases and deaths demonstrate clearly enough that malaria has ceased to be a public health problem in this country. Difficulties have been noted in identifying the results of the campaign because of the prior downward trend. Moreover, evidence has been cited that the number of reported cases is not always a good indicator of prevalence of the disease and that many of the actual cases had been contracted outside the United States. In one study, only slightly more than one-sixth of cases reported in four states (Alabama, Georgia, Mississippi, and South Carolina) were found to be actually malaria, and two-thirds of these had not been contracted in the United States. If this situation were general and typical of all periods, vital statistics on the incidence of malaria might do little more than indicate a relative trend, while failing to present a reliable measure of the absolute magnitude of change in mortality and morbidity from the disease.

The downward trend in the incidence of malaria, before the DDT-based campaign was initiated, has important implications

Table 17. Resources Used in Antimalarial Campaign, United States, 1945–1949, and Extent of Campaign

Fiscal Year	No. of districts covered	No. of houses treated	Quantity of DDT used		Manpower employed
			Total tons	Per house gms	(Man-hours per house)
1944–45	111	264,482	47.2	178	1.75
1945–46	266	1,055,397	324.6	308	1.55
1946–47	297	1,236,841	437.5	354	1.28
1947–48	347	1,374,766	638.9	464	1.36
1948–49	325	1,300,000[a]	680[a]	523	1.5

[a] Approximate.

SOURCE: E. J. Pampana, "Lutte Antipaludique par les Insecticides à Action Remanente," World Health Organization Monograph Series No. 3 (Geneva, 1951), p. 72.

for the model used to assess and evaluate social benefits. These implications will be discussed in the next chapter. The relatively high proportion of cases which, according to some, were contracted outside the United States, may reflect an abnormally high exposure of many American servicemen to malarial infection while serving overseas during World War II. Clearly, however, some social benefits accrue to the United States from programs to eliminate or control malaria overseas.

Important defense implications may be involved. Despite advances in suppressive and therapeutic drugs, the disease is potentially a great incapacitator of armies. In a number of critical campaigns during World War II, success for the Allied troops was threatened by the heavy casualty rates sustained from malaria. The disease is currently a problem to the U.S. forces in Vietnam.

BIOLOGICAL EFFECTS ON MAN

Many of the new pesticide chemicals (like their predecessors—arsenic, nicotine sulphate, and the like) are toxic to man. In some cases their toxicity is many times greater than that of older-type chemicals and, furthermore, they are often much more readily absorbed into the human body. Organic chemicals can enter the body by ingestion, dermal absorption, or inhalation. Oral ingestion presents hazards in two main forms: accidental poisoning, especially of children and irresponsible adults; and involuntary ingestions of toxic residues remaining on or in food that has been treated with pesticides or accidentally exposed to them. Poisoning by the absorption through the skin presents hazards chiefly to those engaged in the production, distribution, and application of pesticides, but also, potentially, to consumers of textiles mothproofed with persistent insecticides, and to members of the public who are exposed to an environment contaminated by chemicals or who happen to be in or near the area being treated at the time. Exposure of agricultural workers in harvesting, processing, or simply working among treated crops has also been demonstrated to present occupational hazards. Dangers of accidental poisoning through the skin also exist, of course, where children have access to chemicals. Inhalation is chiefly an occupational hazard, although incidental exposure of third parties may also occur through this route. The President's Science Advisory Committee [4] has pointed to the possibility of inhalation of airborne pesticides, dust from treated soil, and house dust from mothproofed rugs and blankets.

The extent and nature of these hazards are described in the following review, which also discusses measures taken to modify them.

Occupational Exposure

One observer in 1953 [5] suggested that the chief hazards experienced by persons engaged in the production of pesticides occur not in the manufacture of the basic chemical, but during formulation of pesticides. This process, it was pointed out, is typically carried out by a large number of firms, many of which are small and lack good equipment. The hazards involved are considered amenable to control by ordinary measures of industrial hygiene.

Occupational hazards to agricultural workers and to persons engaged in agricultural spraying and dusting operations are, however, of increasing concern. The Governor's Special Committee in California reported that there had been a three-fold increase in occupational diseases attributed to agricultural chemicals in California during the 5 years ended 1959 [6]. There were 1,100 such cases in California in 1959, mostly among agricultural workers. The dangers of dermal absorption are, perhaps, less obvious than those of other methods of ingestion, and its importance is sometimes overlooked in cases of occupational illness. Hayes [7] cites a report on cases of poisoning of workers among crops treated with parathion. Mild clinical poisoning was observed in 78 of 149 persons involved in 12 separate episodes. In most instances, residues were relatively fresh but, in a few, residues were from 12 to 33 days old. Parathion is generally regarded as one of the most rapidly degraded chemicals in common agricultural use, but others [8] point out that its persistence depends upon the material treated.

Safety records of pesticides among agricultural workers of different countries appear to be related to the general level of literacy and technological sophistication among rural populations. Careful attention to labeling and to printing on the container of full instructions for the use of pesticides, precautions to be observed, and recommended first aid procedures has proved to be one of the most important measures for reducing health hazards. Other possible measures to protect agricultural workers from harmful exposure to pesticides include:

(a) safety education programs among spray or dust operators;
(b) control over the form and containers in which chemicals are distributed (such as, for example, restrictions on marketing of chemicals in concentrated form, where measuring or diluting operations are dangerous, or restrictions on the incorporation of wetting agents in the formulation, where dangers of absorption of chemicals through the skin is high);
(c) direct regulation of the use of pesticides (requiring permits for purchase of designated, highly dangerous chemicals and, perhaps, their application by licensed operators with specialized experience, training, and skills; prescribing the conditions of usage of agricultural chemicals so as to reduce the occupational hazards involved, etc.).

It was stated in 1960 [7] that in only one state in the United States, namely, California, is there any extensive direct regulation of the use of pesticides and that nowhere in this country are regulations for the protection of workers engaged in the application of toxic chemicals developed to such a high degree as in the United Kingdom and some other countries.

Nevertheless, the United States has a relatively good safety record with the newer pesticides, especially with parathion, one of the chemicals for which problems of occupational and accidental poisoning have been greatest. In the period beginning with its commercial introduction, 1947–1960, deaths from parathion in the United States numbered 100. This can be contrasted with 100 fatal cases of parathion poisoning in India each year, 67 annually in Syria, and 20 in Jordan. There were in Japan, for several years, about 1,500 cases of parathion poisoning annually, with an average of 336 deaths per year. The high figures for Japan reflect heavy usage of pesticides in an agriculture characterized by very small holdings. Expenditures on pesticides annually exceed inputs of machinery services in Japanese agriculture. However, Japanese mortality rates may be inflated to some extent by a greater preference for suicide by pesticide than is found in any other country.

Accidental Poisoning

Acute accidental poisoning from pesticides is much greater among very young children than other age groups and most cases are still the result of oral ingestion of older-type pesticides.

Statistics of annual deaths and nonfatal poisoning from pesticides are not separately published, but are recorded under two general categories of accidental deaths caused by "chemicals, solid and liquid" and "chemicals, gases and vapors." Most poisoning by pesticides almost certainly falls within the category of accidental poisoning by solids and liquids, since poisoning by fumigants is believed to form only a small proportion of accidental poisoning by gases and vapors. Accidental deaths in the United States caused by "chemicals, solid and liquid," totaled 2,061 in 1963. On the evidence of previous surveys and analyses, probably about 10 per cent of these were due to pesticides [7]. One study of accidental poisoning in the United States in 1956 concluded that deaths caused by solid and liquid pesticides constituted 9.8 per cent of those caused by poisoning with solid and liquid substances generally. A similar study based on death certificates and correspondence with physicians was done for 1961 [9]. The results of these studies are compared by pesticide compound in Table 18.

In both 1956 and 1961, children under 10 years of age accounted for a large share of the deaths (61 per cent in 1956, 51 per cent in 1961). This is in marked contrast to countries such as

Table 18. Identity of Pesticides Responsible for Accidental Deaths in 1956 and 1961, United States

Pesticide	No. of deaths 1956	No. of deaths 1961
Arsenic	54	29
Phosphorus	21	12
Thallium	8	2
Mercury	1[a]	—
Nicotine	4	3
Strychnine	3	1
Rotenone	1	—
Other inorganic and botanical solid and liquid pesticides	4	11
Subtotal—inorganic and botanical solid and liquid pesticides	96	58
Diazinon	2	1
Demeton	1	—
Malathion	3	3
Methyl Parathion	—	3
Parathion	11	15
TEPP	2	—
Other organic phosphorus insecticides	1	2
Subtotal—organic phosphorus insecticides	20	24
Aldrin	1	—
BHC (including lindane)	2	1
Chlordane	3	—
DDT	1	—
Dieldrin	1	1[a]
Endrin	3	1
Toxaphene	1	2
Combination of chlorinated hydrocarbon insecticides	1	1
Subtotal—chlorinated hydrocarbon insecticides	13	6
Other and unspecified solid and liquid pesticides	16	17 (2[a]) (2[b])
Subtotal—solid and liquid pesticides	145	105
Cyanide gas	1	3 (1[a])
Methyl bromide	3	—
Other gases and vapors (pesticides)	3	3
Subtotal—gases and vapors (pesticides)	7	6
Grand Total	152	111

[a] Diagnosis of pesticide poisoning open to serious question.
[b] Diagnosis of pesticide poisoning open to some question.

SOURCE: Wayland J. Hayes, Jr. and Carl I. Pirkle, "Mortality from Pesticides in 1961," *Archives of Environmental Health,* Vol. 12, January 1966, p. 46.

England and Sweden where only 5.8 and less than 10 per cent of deaths due to poisoning occur among children [9]. In 1956, 68 per cent of the deaths were due to pesticides in use before the introduction of DDT. In 1961, 58 per cent of the deaths from pesticides were due to the older compounds. Roughly equal amounts of about 30 per cent or more were due to the newer synthetic compounds. In comparison, barbituric acid and its derivatives accounted for 323 accidental deaths in 1956.

The study of deaths in 1961 indicated that the deaths among rural people were about the same percentage of total pesticide deaths as the rural population is of the total population. Non-white deaths from pesticides accounted for 30.6 per cent of the total in 1961, while non-white people comprised 11.5 per cent of the population. On a geographical basis, the South Atlantic region, with 14.5 per cent of the population, accounted for 35.2 per cent of the pesticide deaths in 1961. In contrast, the Pacific region where pesticides are used most intensively, accounted for 11.8 per cent of the population, but only 8.1 per cent of the deaths were due to pesticides.

Hayes and Pirkle point out the need for better diagnosis by physicians [9]. In two of the cases in 1961 ascribed to pesticides on the death certificate, it was possible to exclude this possibility. They suggest some misdiagnoses result from failure of the doctor to examine the container. Some physicians may assume all insecticides are DDT.

Information on nonfatal poisoning cases is more fragmentary. Figures compiled by the American Red Cross show that among 2,407 hospitalized poisoning cases the ratio of nonfatal to fatal cases average 50 to 1 [7]. Some 5.8 per cent of the nonfatal cases and 4.3 per cent of the fatal cases were associated with pesticides. The sample in this study was drawn from an urban area, which may explain the lower percentage of fatal cases due to pesticides than figures given above for the entire United States in 1956.

A number of surveys have also been made to find evidence of illness among persons unintentionally or incidentally exposed to pesticides. It has been found possible to measure the excretion of P-nitrophenol in the urine of people whose only exposure to parathion occurred as a result of their living near treated orchards. This is perhaps not surprising when data on the efficiency of application of insecticides is considered. Frequently substantial proportions of chemicals used are deposited downwind as a result of drift or volatilization of chemicals. One study [10] showed that some 45 per cent of chemical spray emitted from an airplane was found on the ground between 10 and 1,000 yards downwind. Deposit densities in this experiment were 3 cc per square yard under the plane and were still as high as 0.1 cc per square yard at points 100 yards downwind. The study also pointed to the low deposits per unit emission in the case of insecticides applied to deciduous trees as fine sprays using ground equipment. It was found that before the trees came into leaf 70 per cent of fine spray was wasted by drifting out of a half-acre plot being treated.

To some, this textual juxtaposition of the presence of a parathion derivative in the urine of bystanders and the typical dispersion of pesticides beyond the target area may be sufficient to inspire logical leaps to gloomy prognostications of insidious self-inflicted poisoning of the race. To dispel most of these fears it can be pointed out that no clinical evidence of illness was discernible in persons suffering incidental exposure to parathion; illness was invariably associated with much higher levels of excretion of the parathion derivative. Nevertheless some feelings of uneasiness may still be experienced outside the medical profession, especially when the long failure of clinical methods to establish relationships between various diseases and sustained cigarette smoking or sustained breathing of polluted air is considered. Clinical methods, that is to say, may well be insufficient of themselves to establish causality in this field of public health.

Toxic Residues in Food

The problem of toxic residues on or in foodstuffs is not newly arisen with the appearance of synthetic organic pesticides. Improper use of older type pesticides has led to illness of food consumers on many occasions. Of the several instances of poisoning from pesticides on foods cited [7], one episode involved the illness of 11 persons who ate mustard greens that a farmer had sprayed with double the recommended dosage of nicotine sulfate and then sold the next day. However, the growing number of chemical formulations (more than 54,000 active formulations were in effect as registered under the federal Insecticide, Fungicide and Rodenticide Act as of mid-1962), together with the greater toxicity and persistence of many of the newer chemicals, has complicated the task of protecting consumers.

Safeguards against harmful effects are provided by a system of tolerances, by testing foodstuffs for the presence of residues, and by education on methods of pesticide usage that will avoid or minimize residues. A tolerance is defined in the World Health Organization principles [11] as the permitted concentration of a pesticide chemical, its derivatives, and adjuvants in or on a food. Tolerances are set after tests of acute and chronic toxicity using several test species of animals, and are usually related to a standard fraction of the minimum amount or concentration of the substance at which harmful consequences become apparent for the most sensitive of the test animals for which observations are obtained. Zero tolerances are mandatory in the United States for known cancer-producing substances under the so-called Delaney clause of the 1958 amendment to the federal Food, Drug and Cosmetic Act.

Responsibility for determining and enforcing tolerances or maximum amounts of residues permitted on foodstuffs is shared between federal and state governments. The federal government sets and enforces tolerances for foodstuffs entering interstate trade. Control of residues on food traded within state boundaries is a matter for state governments.

The principal legal instruments of federal regulation of pesticides are the Insecticide, Fungicide and Rodenticide Act (1947) and the Food, Drug and Cosmetic Act as amended in 1954 (the Miller amendment). Responsibility for administration of these two laws is assigned to the U.S. Department of Agriculture and the U.S. Department of Health, Education and Welfare. The first act provides that pesticides must be registered in order to be shipped legally in interstate commerce. The USDA is responsible for such registration of pesticide compounds.

Prior to May 1964, an applicant denied registration by the USDA could, and on some occasions did, demand that the product be registered "under protest." The product could then be sold in interstate commerce until the USDA could develop evidence to justify a court decision to remove the product from the market. A product, so registered, could be sold until the issue was resolved. This part of the law was changed by Public Law 88–305. A mechanism for appeal from USDA decisions was substituted for the protest registration procedure.

Actually, a relatively small number of products were registered under protest. However, the new law prevents any chance of extremely dangerous situations arising from poorly researched effects, poor labeling or ill-advised use, in addition to preventing the sale of compounds that will not perform as claimed.

Applications for registration of a chemical are accompanied by a detailed petition presenting the results of tests on effectiveness of the chemical for the specific purposes proposed, and its toxicity to mammals. If the preparation is not intended for use on food products, the USDA grants registration, provided it is satisfied that there are not undue hazards to man or domestic animals when the chemical is applied for the particular uses intended and in accordance with the instructions on the label. If use on food crops is proposed, the application must include experimental data concerning the amount and nature of residues. If it is demonstrated that no residue is left on a particular crop when the chemical is used according to instructions, the USDA approves registration. If, however, there is a residue, registration is not normally effected until a tolerance has been established by the Food and Drug Administration.

Tolerances are usually set at 1 per cent of the lowest level at which harmful effects were observed on the most sensitive species of animals used in the tests. This rule of thumb is presumed to provide the necessary safety margin against the possibility of greater sensitivity of human beings than the test animals and of variations in sensitivity among humans of different ages and states of health. When two or more pesticides of similar pharmacologic action are present, the combined fractions of tolerances present should not exceed 1. In 1960 more than 2,000 tolerances involving more than 100 pesticidal chemicals had been established in the United States.

The Miller amendment to the federal Food, Drug and Cosmetic Act assigned to the U.S. Department of Health, Education and Welfare the responsibility for monitoring and enforcing tolerances. The Food and Drug Administration of this department regularly examines samples of food shipped interstate or overseas for the presence of pesticide residues in excess of tolerances. Limitations in staff and laboratory facilities available for this purpose preclude testing of more than an extremely small proportion of food supplies entering international trade.

It seems that only rarely is a sample of food analyzed for which residues exceed tolerances. In New York State in 1962, 1,116 samples were analyzed by the Buffalo Laboratory of the Food and Drug Administration covering 10 varieties of fruit and 31 vegetables and vegetable products [8]. None were found to exceed the legal tolerances.

Seizures by the Food and Drug Administration and tests of foodstuffs by the California State Department of Agriculture can be used to indicate the amount of raw agricultural produce in the United States containing pesticide residues in excess of the tolerance levels [12]. "During 1961 a total of 385 tons of raw agricultural commodities containing unpermitted pesticide chemicals or residues in excess of established tolerance level was seized by the Food and Drug Administration" [12]. Eight of the seizures involved grain contaminated by admixture of seed grain treated with mercurial fungicides. Other seizures included leafy vegetables with illegal residues of parathion, toxaphene, and DDT and peanuts containing excessive amounts of DDT. Fewer seizures were necessary in 1961 than in previous years. In 1960, 60 seizures totaling 1,377 tons were necessary—the most notable concerning cranberries containing residues of the herbicide aminotriazole and apple pomace contaminated with DDT.

The California State Department of Agriculture in 1960 tested 2,166 samples of food for pesticide residues. Of these 141, or 6.5 per cent, were found to contain excessive residues. In comparison,

8.8 per cent of 1,748 samples carried more than the permitted quantities in 1959 and, in 1958, 8.3 per cent of 1,846 samples. These percentages should not be equated with the percentage of total supplies of raw agricultural produce carrying illegal residues. The department emphasizes that the proportion of positive test findings is greater than would be found by random sampling, since inspectors selectively sample commodities suspected, for some reason, of carrying excess residues.

During the fiscal year ended June 30, 1964, the Food and Drug Administration instituted 42 seizures of agricultural commodities nationwide. Of these, 34 were raw commodities. All were vegetables except 18 seizures which included eggs, alfalfa hay, wheat, and barley. The barley and wheat carried mercurial fungicide and the others carried chlorinated hydrocarbons [13]. In west Texas, seizures disclosed evidence of drift of chemicals during application; commodities that matured early and, therefore, enough time had not elapsed between application and harvest; endrin take-up from the soil by carrots; deliberate misuse; and carelessness and ignorance (benzene hexachloride was mistaken for fertilizer). The Food and Drug Administration samples 1 per cent of the interstate shipments on a surveillance basis with no *a priori* knowledge. Additional lots are sampled where residues are suspected.

There are no recorded instances of seizure of crop following the use of pesticides in accordance with instructions. Usually excess residues have followed gross disregard of the recommended practices regarding the amount used and/or timing of the application.

In most states, legislative provisions similar to or identical with those at the federal level are in force to control the distribution of pesticides and to protect the public against toxic residues in food. In May 1963, 43 of 50 states had regulatory programs providing for registration of pesticides, based on standards of effectiveness and safety. Moreover, more than three-fourths of the states were reported by the Wisconsin Governor's Special Committee [14] to have food laws which are, in essential respects, identical to the consumer protection provisions in the federal populations. "We are rightly appalled by the genetic effects of radiation," says Rachel Carson; "how, then, can we be indifferent to the same effect in chemicals that we disseminate widely in our environment?" [16]. There has also been some speculation that continued ingestion of small quantities of pesticides could result in serious, but insidious, somatic effects upon more sensitive individuals within a population. Such effects, it is held, might not become apparent until widespread irreversible damage has been done.

Present public policy in the United States is somewhat inconsistent in its approach to such secular or speculative effects of pesticides on health. Only in the case of carcinogens is the no-threshold hypothesis accepted. The Delaney clause of the 1958 amendment to the federal Food, Drug and Cosmetic Act provides that no carcinogenic (cancer-producing) substance be permitted in food in any quantity (incidentally an extremely difficult requirement to administer, since widely encountered ingredients of food such as common salt and even food itself in sufficient quantities are known to be carcinogenic). It is, of course, conceivable that different hypotheses might be valid for mutation-inducing or neurotoxic substances and for carcinogens, but there would appear to be no empirical evidence to support this. In any event, imposing a zero tolerance simply because one type of risk is involved, while nonzero tolerances are set on other types of health risk, regardless of the magnitude of the risk and of the outcome, is patently inconsistent. If the no-threshold hypothesis of dosage-effect relationships held good in both cases, the decision as to tolerance or permissible levels would become an economic decision involving the weighing of costs and probabilities of deleterious consequences against the benefits derived.

If ridiculous decisions are to be avoided, however, the need for regard to the quantitative relationship strongly favors retention of the threshold hypothesis. Naturally occurring toxic substances are found in many foods. Durham [12] mentions the oxalate in cabbage, spinach, and rhubarb and observes that Swedish turnips may be 100 times more potent with regard to antithyroid activity than cranberries badly contaminated with aminotriazole. The

application of a no-threshold hypothesis to these naturally occurring constituents of food would bar the consumption of
vitamins A and D, cobalt, and fluorine—compounds beneficial in
small quantities, but toxic in large amounts.

Interaction Hazards

One of the possibilities that concerns thoughtful people is
synergism, or the interaction of chemicals. Synergism is the production of an effect by the combination of two or more chemicals that is greater than would be expected on the basis of a
simple additive effect. Durham [12] notes that various combinations of the organic phosphorous compounds have been shown
to potentiate or interact with each other. The greatest observed
synergistic effect was obtained by combining malathion with
triorthocresyl phosphate (TOCP). The combined toxicity of
these two chemicals was from 88 to 134 times greater than their
effect administered singly. The mechanism of potentiation appears to involve interference of one compound with the metab-
olism of the other. Thus, the enzymes that normally enable man
to convert malathion to relatively harmless materials are blocked
[7].

Although it is said that it is unnecessary to invoke potentiation
to explain any case of human poisoning—even occupational
poisoning—the possibility of conjunctural combinations of chemicals leading to synergistic effects upon man remains a matter of
concern.

Some observers point to the very considerable research effort
to achieve more effective pesticides by potentiation. They regard
the small number of cases discovered as sufficient evidence that
the risks of accidental synergism, while perhaps warranting care
in occupational exposure, should not occasion serious concern in
the case of incidental exposure or toxic residues on food. Some
have taken the view that harmful effects of synergism are at
least as remote a possibility in pesticide usage as in other fields in
which they are virtually disregarded in decision making, or have
only been recognized recently, e.g., water and atmospheric

6

Evaluating the Effects of Pesticides on Health

We discussed in the preceding chapter the form and incidence of the effects of pesticides on the health of human populations. We now turn to an examination of the problem of measuring and evaluating these effects for the purpose of incorporation into a benefit–cost analysis.

NONMARKET OBJECTIVES

There is some disagreement among benefit–cost analysts concerning methods of bringing nonmarket objectives, such as health, into a decision-making model. Some argue that, with sufficient ingenuity and persistence, ways can be found to express health effects in terms of some common denominator that allows them to be considered in conjunction with other (market) objectives. Others take the position that the tenuous calculations and the mental gymnastics required to translate "incommensurable" factors into money terms should be avoided. They suggest instead that the best way to deal with "incommensurables" is to append for the information of the decision maker a discussion in qualitative or, if possible, quantitative, terms to the main analysis. This procedure avoids ignoring important elements of the problem but does not attempt to do so by forcing multiple objectives into an analysis based on a single or monetary objective. It thus avoids overstraining the data with heroic valuations and avoids imparting a possibly spurious air of precision to essentially imprecise calculations.

A third treatment of nonmarket objectives is the use of restrictions to represent the nonmarket objective, thereby limiting the attainment of market objectives expressed by measures such as per capita income, etc. This general valuation approach was mentioned in Chapter 2. Measures of the sacrifices in market value objectives required to increase the level of the nonmarket objective can be obtained.

Suppose, for illustrative purposes only, that a restriction on the use of a set of pesticide compounds increases the cost of the annual national food bill by $4 billion for a constant amount of food of constant nutritional quality. Suppose also that this action is calculated to save, over a 20-year period, 1 million lives. This amounts to an annual outlay of $4 thousand per life saved or a sacrifice in the satisfaction from other goods and services with a value equal to $4 thousand. This is the amount to ensure that one life will be protected from the acute or chronic effects resulting from the ingestion or exposure to this set of pesticide compounds. In this hypothetical example, the opportunity cost of saving 1 million lives is $4 billion per year. To adopt the policy of restricting the use of this set of pesticide compounds is to say, in effect, that the 1 million lives are worth an amount equal to or greater than $4 billion annually. To refuse to adopt such a policy is to say, if the decision is rational, that $4 billion per year for 20 years is too much to pay to save 1 million lives.

Obviously, there are questions of risk and uncertainty that would enter any such actual calculation. Such a calculation nevertheless does, if performed for a number of courses of alternative actions, provide the relative opportunity costs of fulfilling certain nonmonetary goals. The importance of such opportunity cost information is made apparent when public decisions and their effects are viewed, not from the standpoint of developing least cost means of preventing death, injury, or sickness from floods, pesticides, or automobiles individually, but rather when some combination of actions on all hazards is considered so that the opportunity cost of preventing a death due to pesticides is just equal to the cost of preventing a death due to automobile accidents, etc. In this way, there is less chance that undue concentration will be given to improving health by one means when another offers equal or greater improvement for less sacrifice.

It has been claimed that values for nonmarket objectives are implied in past decisions and that by comparing the actual course of action selected with the optimum that would have been feasible in the absence of imposed constraints, the difference can be taken as a measure of the value the decision maker has placed upon the nonmarket objective. Thus social values attached to, say, health objectives could be derived by calculating the sacrifices in the level of achievement of primary objectives such as, perhaps, additional output and resource savings in agriculture. Though fraught with difficulty, this approach can lead to rather important conclusions concerning the rationality of environmental standards set in the interest of health and safety. The difficulties are discussed below.

Government Decision Making

The limitations imposed by partitioning of government decision making among bureaus with specific functions have been noted by Margolis [1]. The government, he observed, does not constantly consider all alternatives. It lacks a well articulated preference function and an ability to consider all aspects of the economy. The frequent result, therefore, is a neglect of goals more general than those of an agency or of alternative programs (offering cheaper means) to achieve a given goal. Hitch and McKean issue similar warnings against the use of implicit values from historical decisions for valuing human life in defense planning. They note that the values attached to human lives "as implied in World War II decisions, differed from one situation to the next and were not necessarily the values that should then or ought now to be attached to lives" [2]. There is no well-organized quasi-market for public health and other nonmarket objectives, if by quasi-market we mean an area in which government institutions and the planning process have tended to bring decision-implied prices or values for a given objective substantially into equality.

Benefits and Costs

In order to ensure that the effects of pesticides upon health are sufficiently recognized, however, some attempt to derive money values for health benefits or losses seems desirable. Confining their treatment in, or rather their bearing on, the analysis to a supplementary note could easily submerge their importance.

The two chief barriers to evaluation of the effects of pesticides through reduced mortality and morbidity from disease consist, first, of inadequacies of statistical data concerning the incidence of disease and the costs of illness and, second, lack of knowledge concerning the functional relationship between pesticide usage and health. The former deficiency concerns the lack or inadequacy of separately recorded statistics for pesticide-related diseases of the age, sex, and number of persons affected, duration of absence from employment, costs of medical care, and percentage debilitation as a result of infection. Similar but somewhat less formidable difficulties complicate the task of assessing the health effects of pesticides through accidental, occupational, and incidental poisoning and of incorporating these consequences into the decision-making model for public policy in pesticides.

It is not possible, in the present study, to overcome these difficulties and to present quantitative estimates of the benefits and costs of pesticides to public health. However, it is possible to indicate the components of benefits and costs and to draw conclusions concerning the research and data-collection needs for further progress in economic appraisal.

Conceptually, the same model as outlined in Chapter 3 applies to the effects of pesticides on human health. The health benefits (or costs) of pesticides, like their effects on agricultural production, will be manifested both through reduced (or increased) mortality and morbidity and through reduced (or increased) resource requirements.

The concepts of demand and supply curves for human lives seem somewhat distasteful and, in any case, difficult to give empirical content. Clearly, however, society will be prepared to save fewer lives or to prevent fewer cases of disease as the cost

Food, Drug and Cosmetic Act. However, the same report claimed that the Wisconsin State Department of Agriculture was (in 1960) the only agency other than the federal Food and Drug Administration testing food products other than milk for pesticide residues. This claim is incorrect, but would suggest that foods confined to intrastate trade are not subject to the same degree of testing to ensure conformance to tolerances as are foods entering interstate commerce. A considerable part of the nation's food may therefore currently escape monitoring programs.

Hypotheses Underlying Tolerances

The concept of tolerances or acceptable levels of contamination of foodstuffs, like the concept of selective toxicity, implies acceptance of the threshold hypothesis. That is to say, for all chemicals there is some level of dosage or concentration, greater than zero, at which there will be no (harmful) effects on exposed individuals. The same concept of the tolerance dose was formerly used in the development of radiation protection standards. In recent years, the accumulation of evidence has cast doubt on the assumption that there is a positive dose that will be safe for all exposed individuals. Evidence on the genetic effects of radiation indicates that extremely small doses delivered to the gonads prior to reproduction will be accompanied by an increase in genetic mutations, most of which are deleterious [15]. The threshold hypothesis has therefore been discarded and an alternative hypothesis accepted that any dose is accompanied by an increased risk of deleterious biological effects, the magnitude of the risk increasing with the dose. The establishment of standards for radiation protection involves balancing the risks inherent in a particular level of exposure against the benefits to be derived.

The no-threshold hypothesis of dose-effect relationships has also been advanced recently for pesticidal residues. While there is not an accumulation of supporting evidence comparable to that for radionuclides, some observers have pointed to the possibility that genetic or other insidious damage may result from ingestion of chemicals in trace amounts by humans or wildlife pollution or drug merchandising. If a real risk of potentiation exists, however, it should be considered in the decision concerning pesticides. The fact that decisions may customarily be made on the basis of incomplete information in other areas of public policy does not constitute a valid argument for excluding the probability of synergism, however small, from the decision-making model.

By its very nature, synergism or potentiation constitutes an uncertainty and not a measurable risk. Even if observations over time would eventually yield meaningful estimates of risks, it is unlikely that this way to their measurement would be left open. This is especially true if the risk to be assessed is not the chance of a few thousand deaths or obvious cases of genetic mutation (observed in newborn infants with every fiftieth new basic pesticidal chemical or every ten-thousandth new commercial formulation marketed), but whether in the long run we shall all be dead of synergistic poisoning.

Conceptually, the degree of uncertainty concerning synergism can be moderated by public action. Dangers of discovering synergism after a tragedy could be reduced, for example, by requiring toxicity tests of new chemicals in combination with other common pesticidal ingredients as well as singly or by making compulsory commercial trials of new chemicals in a limited geographical region prior to their general release. Such tests would inevitably be very costly in view of the large scale of experiments necessarily involved. They might, moreover, be of limited usefulness in view of the extremely large number of possible combinations to be eliminated for each chemical and the possibility of apparently nonpotentiating combinations becoming activated only under certain conditions.

CONCLUSION

After a detailed review of medical knowledge concerning potential health hazards of pesticide residues in food and water, one scientist concludes that there are indications that the residues do not pose a significant threat to human health at the present time [12]. There has been, for example, no progression in the average level of pesticide residue in the general food supply of the United States and no significant change in storage of DDT in the general population between 1950 and 1962. However, continued research on the effects of subacute dosages on humans is warranted.

We are warned, however, that the possibility of qualitative differences between effects on man and effects on test animals cannot be ruled out and that, for this reason, it is highly important for a close surveillance to be maintained of workers who handle pesticidal formulations in agriculture and public health work. Even with the best available safety precautions, the exposure of these workers is far greater than that of the general population and any subtle or delayed effects would occur first and perhaps exclusively among them, giving a warning of the slightest danger to the general population.

The conceptual basis of tolerance appears to be sound and consistent with knowledge of the etiology of pesticidal compounds. Perhaps the main questions for resolution, insofar as the health hazards of pesticides are concerned, are the adequacy of safeguards against occupational disease among workers handling chemicals, the adequacy of food-testing facilities especially for food not entering interstate trade, and the possible need for further research on pesticide residues in water that have been studied rather less intensively than residues on raw agricultural produce.

REFERENCES

1. S. W. Simmons, DDT, *The Insecticide Dichlorodiphenyltrichloroethane and Its Significance* (Basel, Switzerland: Birkhauser Verlag, 1950).
2. E. J. Pampana, *Lutte Antipaludique par les Insecticides a Action Remanente*, World Health Organization Monograph Series No. 3 (Geneva, 1951), p. 72.
3. L. L. Wiliams, "Economic Importance of Malaria Control," *Proceedings of the Twenty-fifth Annual Meeting of the New Jersey Mosquito Extermination Association*, Atlantic City, March 24, 1938, pp. 148–51.
4. President's Science Advisory Committee, *Report on Use of Pesticides*, May 15, 1963 (Washington, D.C.: Government Printing Office, 1963), p. 25.
5. J. M. Barnes, *Toxic Hazards of Certain Pesticides to Man*, World Health

Organization Monograph Series No. 16 (Geneva, 1953), p. 129.

6. Governor Edmund G. Brown's Special Committee on Public Policy Regarding Agricultural Chemicals, *Report on Agricultural Chemicals and Recommendations for Public Policy* (Sacramento, California), December 30, 1960, p. 26.

7. W. J. Hayes, Jr., "Pesticides in Relation to Public Health," *Annual Review of Entomology*, Vol. 5, 1960, pp. 379–404.

8. E. H. Smith, *Statement Before Subcommittee on Reorganization and International Organizations of the Senate Committee on Government Operations*, August 21, 1963. Mimeo.

9. Wayland J. Hayes, Jr. and Carl I. Pirkle, "Mortality from Pesticides in 1961," *Archives of Environmental Health*, Vol. 12, January 1966, pp. 43–55.

10. R. J. Courshee, "Some Aspects of the Application of Insecticides," *Annual Review of Entomology*, Vol. 5, 1960, pp. 327–52.

11. World Health Organization, *Principles Governing Consumer Safety in Relation to Pesticide Residues*, Technical Report Series No. 240 (Geneva, 1962).

12. William F. Durham, "Pesticide Residues in Foods in Relation to Human Health," *Residue Reviews*, Vol. 4, 1963, pp. 33–81.

13. Sam D. Fine, "Pesticide Residue Problems in Texas—A Progress Report," *Water for Texas, Proceedings of the Ninth Annual Conference, November 23–24, 1964, Texas A. & M. University*, pp. 57–59.

14. Governor's Special Committee on Chemicals and Health Hazards, *Report on Food and Feed Additives and Pesticides* (Madison, Wisconsin), April 1960, pp. iii, 40. Mimeo.

15. D. R. Chadwick and C. P. Straub, "Guidelines for Tolerance Levels of Radionuclides in Man," *Proceedings of the North Central Experiment Stations Workshop on Radionuclides in Foods and Agricultural Products, Cincinnati, Ohio, February 19–21, 1963*, pp. 65–75.

16. Rachel Carson, *Silent Spring* (Boston: Houghton Mifflin Co., 1962), p. 37.

of saving a life increases. The relationship between the price of a life saved and the number of lives that society will stand ready to save, will, therefore, be characterized by the normal negatively sloped curve. Similarly the supply function—the number of lives that can be saved with the quantity of resources commanded at each level of value placed on a human life—will be positively sloped. In short, lives can be saved at a price and the higher the price placed on them by society the more lives can be saved. Moreover, an innovation in the control of a disease or prevention of accidents will shift the supply curve to the right. The conceptual apparatus developed for measuring social benefits of innovations or of alternative policies in agriculture thus holds equally well for the effects on public health.

The practical consequence of this is that the benefits of a health program cannot be assessed simply by counting the lives or working days saved and putting a value on them. Whether a valuation of human lives (or working time) so obtained would approximate the social benefits of the program would depend upon whether resource-saving consequences of the program are important.

Costs of Disease and Injury

Mushkin and Collings [3] suggest two categories of costs of disease and injury that apply to social costs. The first category is costs of resource use, covering the use of resources for the prevention, diagnosis, and the treatment and rehabilitation of persons infected or injured. The second category of resource loss covers death, disability, and debility, but the concept is not necessarily limited to human resources.

It is possible that the downward trend in incidence of malaria, for example, might well have been maintained after 1947 by increased quantities of conventional antimalarial inputs. It is likely, therefore, that a substantial part of the effects of insecticides used in antimalarial operations accrued to society not as lower incidence of the disease but as lowered resource requirements for its control.

Moreover, underdevelopment and underutilization of natural

resources in the endemic regions may well have been one of the largest resource loss costs. This loss can be included in a conceptual framework based on demand and supply schedules for conservation of human lives and health, if we allow it to be reflected in the social demand curve for control of a particular disease. Society should be willing to pay a higher price for a given number of lives saved (or for reduced incidence of a disease), if by so doing it will open up nonlabor resources previously rendered inaccessible or costly to employ because of the presence of disease. This involves having different demand curves for human resources saved from malaria and, say, cancer, with a given reduction in mortality from malaria commanding a higher price than the same absolute reduction in deaths caused by cancer. This makes empirical application of the model only slightly more difficult. Values placed by society on reduced mortality from various causes will probably diverge, in any event, because of the different age distributions of persons infected with each disease. Moreover, the spatial implications of diseases, and especially their effects on regional economic development, should not be overlooked.

The evaluation of the costs and benefits of health programs (or of innovations in disease prevention and treatment or spillover effects of programs and developments in other fields) can conveniently be divided into three major areas: evaluation of reductions in mortality; evaluation of reduced nonfatal morbidity; and evaluation of change in the quantity and value of resources used in the maintenance of health, valued at opportunity cost.

Valuing Reduced Mortality

It follows from the principle of opportunity costs that human life cannot legitimately be given a supreme or infinite value in a decision-making model. Imposing paramount valuations upon the avoidance of deaths resulting from pesticides (or upon reduced mortality from insect-borne diseases) would clearly lead to diseconomies, possibly resulting in considerably greater mortality from other causes. It could lead to a suboptimal allocation of resources, even if output were measured solely in terms of human mortality.

Various methods for determining the social value of a human life have been employed or proposed from time to time. The use of values implied in previous decisions has already been mentioned and its limitations discussed. The costs of rearing a child to productive age has also been used as a measure of the value of a man. Another method used for evaluating intangibles in a benefit–cost analysis that has been proposed for valuing a human life is the cost of saving a life by the cheapest alternative method. This procedure assumes in effect that the government planning process operates as a price-equating, quasi-market mechanism— an assumption that is patently unjustified in the present state of cross-level, cross-functional, and cross-regional co-ordination within government. When this method was proposed, Margolis [1] thought that it involved the danger of overstating benefits. This fear seems rather ill-founded, however, in the present instance. Indeed the opposite danger would appear more likely, namely that there will be ways open to save lives at bargain prices. Thus, the cost of erecting a stop sign or traffic light outside a school or of a few inches of insulation tape on the worn lead of an electrical home appliance bears a relationship to the value of a human life only in that the cost is considerably less than the value.

Probably the most widely accepted basis of valuing human life is the use of productive value. In this method a human life is valued in terms of the present value of the future labor product of an average person of that age and sex. This method involves the estimation of the stream of earnings of individuals of various ages and sexes over their expected lifetime and the discounting of this stream to obtain the present value of the earnings. If a person's earnings are reduced because of debility, or cut short due to death, the difference in the present value of normal earnings and the present value of actual lifetime earnings becomes an estimate of the social cost of the loss of this life or a fraction of it.

While this method is fraught with difficulties—such as the propriety of gross or net earnings as a measure of productivity, the selection of an appropriate discount rate, the treatment of un-employment and retirement, and productivity not measured by the market, to say nothing of the interpersonal intrinsic value of people—it does bring some order out of a chaotic question.

Discounting the flow of future earnings of an individual to measure the value of an investment in health is necessitated by the opportunity cost principle mentioned earlier. The only and sufficient justification for discounting is the possibility of investing in some other productive investment instead of in public health programs; to ignore time in the decision model could well be to settle for less than an optimal outcome. Most studies to determine the value of investments in the human agent (or to measure the economic costs of a disease or injury) do use discounting to derive present values. An exception is the RAND Corporation study on ulcers [4].

In calculating the present value of future productive services from an individual member of society, we are not implying that a human life can be valued solely in terms of productive capacity. However, such calculations may well help put health problems in perspective and assist in selecting among alternative health pro-grams. As ultimately the value placed on human lives is a political decision (and not yet determinable by reference to quasi-market quotations from previous political decisions), the use of con-strained suboptima to give decision makers an estimate of the implicit values placed on a human life by alternative courses of action is considered to be a potentially useful device.

We have presented an example of this technique in Table 19. While these data do not show losses due to debility or costs of treatment, they do give a relative picture of the loss of the stocks of human capital from various causes of death. As such they pro-vide guidance for investment in treatment, research, and educa-tion related to various causes of death. If, as some suggest, some pesticides are related to the incidence of cancer, then an estimate of part of the social costs of pesticides could be developed by determining that fraction of cancer due to pesticides. Then the question could be approached by deciding whether it is cheaper to cure cancer or prevent it by changing the form of pest control.

Table 19. Estimated Losses of Human Capital (Present Value of Future Earnings) from Selected Causes of Death, United States

Cause of death and year	Discount rate per cent	Aggregate loss (millions of dollars)
Malaria		
1935	4	92.6[a]
1935	10	42.2[a]
Poliomyelitis		
1954	4	44.0[a]
Tuberculosis		
1954	4	253.0[a]
Cancer		
1954	4	1,568.0[a]
Ulcers		
1956	0	257.4[b]
Mental Illness		
1954	4	1,926.5[c]

[a] Estimates computed for malaria by the authors using assumptions given by Burton A. Weisbrod, *Economics of Public Health, Measuring the Economic Impact of Diseases* (Philadelphia: University of Pennsylvania Press, 1961), pp. xv, 127. Estimates for polio, tuberculosis, and cancer computed by Weisbrod. All measurements in 1954 dollars.

[b] Estimated by I. S. Blumenthal, *Research and the Ulcer Problem*, RAND Corporation Report R. 336–R6, June 1959.

[c] Estimated by Rashi Fein, *Economics of Mental Illness*, Joint Commission on Mental Illness and Health Monograph Series No. 2 (New York: Basic Books, Inc., 1958). Covers first admissions to prolonged-care hospitals—not deaths.

In addition, these costs could be weighed off against preventing death and debility from other causes.

Measuring Effects of Debility

Perhaps the most difficult tasks are the measurement of the effects of disease in reducing the vigor, initiative, and productivity of the exposed population and the evaluation of the effects of investment in health programs (or of innovations in control methods) upon debilitation of human resources. Yet in some cases, such as malaria, the effects of debility may far exceed the other costs.

The usual methods of assessing the effects of debility on the efficiency of workers have been summarized [5] and are based on comparisons of:

(a) output of a sample of industrial plants or firms before and after the implementation of control measures;

(b) piece-work earnings of persons suffering from and persons free from the disease;

(c) wages or average earnings in an area where a disease is prevalent and an area in which it is either absent or unimportant;

(d) output in a firm or group of firms carrying out control measures and output of a firm or group not engaging in control;

(e) laboratory tests of work energy or capacity of persons afflicted with and free from a disease.

The percentage reduction in productivity so obtained is applied to total output or earnings of a region to determine the costs of debility from the disease. At times generalizations about debility costs are made on the basis of very restricted samples. When observations are scarce, there is a natural tendency to overwork the ones we have. Thus calculations of the effects of disease through debility have sometimes been based upon a single comparison for a particular type of firm or upon tests conducted with respect to some fairly straightforward and standardized operation [6]. Nowhere in the literature does the concept of the incremental costs of debility or the incremental benefits of reductions in debility appear.

SUMMARY

Evaluation of the effects of pesticides on human health presents a number of complex problems. First, data relating the use of pesticides to health are not currently available in sufficient detail to allow a meaningful analysis of debility and death resulting from pesticides. Second, the placing of values on human life poses serious problems concerning the treatment of unemployment, retirement, and the extramarket values associated with human life and health. Third, the social costs and benefits associated with health programs in general have not been estimated as carefully and completely as they might have been. This latter point raises questions concerning the allocation of resources among alternative health programs, in addition to the problem of resource allocation among health programs and other activities in the society.

The use of past decisions was discussed as a means of placing values on health and human life through the use of values-implied decisions. Use of the present value of future earnings was also discussed as an alternative means of developing estimates of the social cost of disease, death, and injury. The present value method was related to the cost of treatment and prevention as a means of determining the optimum level of health investment similar to the methods used to assess the effects of pesticide technology on agriculture.

Rigorous analysis of the effects of pesticides on health first awaits the development of functional relationships between pesticide usage and pesticide intake by humans and the subsequent functional relationships between pesticide intake and levels of mortality and morbidity. The second step is the translation of the aforementioned relationships into relative losses of human capital, resource costs for treatment, and productivity losses due to debility for the various health effects through the application of economic analysis. Third, the costs of ill health due to pesticides will need to be compared with costs of prevention and/or cures. In addition, costs of preventing or curing ill health due to pesticides will need to be compared with costs of preventing similar sacrifices in health from other causes such as heart disease, auto accidents, etc., which are independent of pesticides.

REFERENCES

1. Julius Margolis, "The Economic Evaluation of Federal. Water Resource Development," *American Economic Review,* Vol. 49, No. 1, March 1959, pp. 96–111.
2. C. J. Hitch and R. N. McKean, *The Economics of Defense in a Nuclear Age* (Cambridge: Harvard University Press, 1960), pp. 185 ff.
3. Selma J. Mushkin and Francis D'A. Collings, "Economic Costs of Disease and Injury. A Review of Concepts," *Public Health Reports,* Vol. 74, No. 9, September 1939, pp. 795–809.

4. I. S. Blumenthal, *Research and the Ulcer Problem*, RAND Corporation Report R. 336–RC, June 1959.
5. Selma J. Mushkin, "Health as an Investment," *Journal of Political Economy*, Vol. 70, No. 5, Part 2, October 1962.
6. Research Council for Economic Security, *A Study of Prolonged Illness Absenteeism* (Chicago, 1957).

7

Effects on Fish and Wildlife

Pesticides have been increasingly used in fish and wildlife management. Clearing of rough fish from lakes, control of the sea lamprey that had decimated the lake trout population of the Great Lakes, and control of rodents and predatory animals are some examples of the use of chemicals in conservation of fish and wildlife resources. Pesticides have also often benefited bird and wildlife populations. Many avian diseases, for example, are transmitted by mosquitoes, and insect extermination or control programs have frequently conferred health benefits on birds and animals. Moreover, some kinds of faunal displacement due to pesticides may lead to more numerous or more vigorous wildlife populations.

However, by no means are all of the effects of pesticides on fish and wildlife favorable. Many pesticidal chemicals are highly toxic to animals, and their use in agriculture, residential areas, and forestry, and in special programs to eliminate newly imported pests, has in a number of instances occasioned adverse spillover effects. Heavy mortalities of fish and birds have followed some programs as a result of biological accumulation of toxic substances and of unforeseen and rather complex food-chain relationships. These incidents, together with the widespread pollution of the environment by persistent pesticides evidenced in traces of the chemicals found in organs of birds and animals, have led to fears that the introduction of persistent and highly toxic chemicals into the environment may disturb complex ecological relationships and result in widespread and irreversible damage to wildlife populations.

Arising out of these fears, pressures have built up for the use of materials which are less toxic to vertebrates or more quickly degraded in the environment. Pressures have also been somewhat reinforced for the use of biological or other nonchemical methods of insect and noxious weed control arising from concern with the human health implications of toxic residues on foodstuffs. The proponents of these views have not always appreciated that there is an economic choice involved in such decisions. Chemicals, relatively nontoxic to vertebrates or mammals, may by virtue of broader spectrum toxicity among arthropods have greater adverse effects on beneficial insects. Additional costs of alternative controls of pests (or of noncontrol) may exceed the benefits to society accruing from reduced losses of fish and wildlife resources. In order for society to make informed decisions when confronted by such choices, it is necessary to have some evaluation of the effects of pesticides on wildlife and some method of gauging the hazards involved in alternative courses of action. We therefore review briefly in this chapter the current knowledge of the nature and incidence of the effects of pesticides on wildlife. The problems of measurement of benefits and costs in this field are then discussed and some conclusions about research needs are formulated. The main sources of information are the National Academy of Science's reports on Pest Control and Wildlife Relationship [1], [2]; Circulars 84 and 167 of the U.S. Fish and Wildlife Service [3], [4]; and the Report of the President's Science Advisory Committee [5].

Nature and Incidence of Effects

In general, pesticide chemicals are not specific poisons for the particular pests they are used to control, but are also toxic in varying degrees to other animals and man. Crustaceans are among the most sensitive organisms. Fish are generally very sensitive to pesticides, especially to endrin and toxaphene. Reptiles, birds, and mammals follow in order of declining sensitivity. Resistant organisms or insect species may concentrate and store pesticides, many of which are fat soluble. At Clear Lake in California, 1,1-dichloro-2,2 (p-chlorophenyl) ethane (DDD) was used to control gnats. The Report of the President's Science Advisory Committee [5] states that although the water of the lake contained only 0.02 parts per million of the insecticide, plankton contained five parts per million. A further concentration occurred in fish where amounts ranging from hundreds to thousands of parts per million were stored in fat. Grebes which fed on the fish died.

A similar biological buildup is believed to have occurred when the robin population in a midwestern city declined after feeding on earthworms, resistant to DDT, which had fed on fallen elm leaves from trees treated with this chemical. Also, aquatic plants have built up concentrations of pesticides from water containing minute traces of chemicals. The complexity of these relationships, unforeseen before heavy mortality among birds or fish was observed, makes some observers extremely unhappy over man's lack of humility in tampering with the balance of nature. They consider that the complexity of the ecosystem speaks poorly for the outcome of man's intervention. Others consider that a few disasters among wildlife, such as at Clear Lake, California, are part of the learning process and that no one could reasonably have been expected to foresee this outcome when the chemical was applied.

Another phenomenon arising from the storage of toxic residues in fat has also occasioned concern. This is the fact that sublethal concentrations may become lethal if fat deposits containing concentrations of toxic chemicals are metabolized during periods of stress such as illness or restricted diets. Poisoning of robins has occurred from DDT stored in fat deposits when the birds were called upon to utilize these reserves [6]. In the conclusion to an introduction to the review of research on pesticide–wildlife investigations in 1961–62, it was noted that further research was required on the "effects of sublethal levels of pesticides, particularly combined with factors of stress such as disease and starvation" [4].

Other sublethal effects on which knowledge is still inadequate include the effects of pesticides upon the fecundity of birds and other wildlife species. Wildlife biologists and bird lovers fear that successful nesting, measured by the number of young birds reared, will be reduced either by effects on the fertility and virility of adult birds or by virtue of the greater sensitivity of the young to poisons.

A number of studies have implicated pesticides (particularly endrin, toxaphene, and rotenone) in fish kills and in lower fecundity of fish populations. A recent report from workers at Cornell University reveals that DDT accumulations in lake trout reduce their reproductive rate [7]. The U.S. Department of Health, Education and Welfare has attributed fish kills in the lower Mississippi River to endrin which, like toxaphene, is toxic to fish at concentrations of less than one part per billion. It is possible, although somewhat unlikely on the basis of earlier studies, that concentrations of the chemical sufficient to cause large-scale fish kills have resulted from run-off of endrin applied in normal agricultural usage.

A study of the pesticide content of a river system in Alabama, draining a cotton-growing area of about 400 square miles, showed that the pesticides DDT, toxaphene, and lindane were extensively used throughout the area. Toxaphene and lindane were detected in the stream but no effects on aquatic life were observed [8].

Nevertheless, as noted above, the hazards of pesticides in water have received somewhat less attention in research programs than residues in food. It is, therefore, reassuring to note that the U.S. Department of Agriculture has undertaken a detailed study to ascertain and to monitor the levels of pesticides in water, soils, and other parts of the agricultural environment in the lower Mississippi River area, as well as in other areas throughout the country.

It is difficult with present data to gauge what effects the widespread use of insecticides and pesticides have had and are having on wildlife populations. Few surveys or censuses of wildlife have been carried out.

Bird Populations

Since bird counts extending backwards for more than 60 years are known to have been made and recorded, it would seem that some measure of trends in bird numbers should be obtainable. The Christmas bird counts of the National Audubon Society are, in fact, of limited value in establishing trends in bird population and distribution, since the number of observers, although known, is not constant, and the counts are taken each winter when many birds have migrated south [9]. The counts are essentially number-counting outings by amateur bird watchers and through the years better places to spot birds have been found. Some scientists are disturbed at the way the counts are being used as observations to determine trends in populations and, in particular, at the conclusion drawn from increasing counts that pesticides are having no major or permanent impact on bird populations.

One must agree that the Audubon Society bird counts are an end in themselves and not a means to describe and analyze changes in bird populations. Although counting has been a highly organized activity for several decades and records of observations of nearly 500 species have been published for each district, there have been very few attempts to interpret these observations. Indeed, only since 1949 have totals for each species observed been published in the *Audubon Field Notes.*

A more systematic count, based on a purposively sampled selection of types of habitat, is conducted by the Audubon Society in co-operation with the U.S. Fish and Wildlife Service. This is the winter nesting bird census. Apparently no published statistical analysis of the results of this count has been made. However, the data have been gathered and could perhaps be analyzed to obtain measures of changing densities and geographical distributions of bird populations.

Even if we knew the magnitude and direction of changes in bird numbers, however, we should still face great difficulties in estimating the effects of pesticides. For one thing, changes in bird populations are a result of a complex of environmental and other factors, including encroachment on habitats from urbanization, highway development, drainage and water conservation projects, land clearing, stream and atmospheric pollution resulting from industrialization, and from increasing densities of human settlement. Moreover, weather and related phenomena, such as grass and forest fires, probably cause considerable fluctuations in numbers, which may also possibly exhibit cyclical tendencies as a result either of inherent characteristics of the species or of cycling in some part of the food chain.

Probably aggregate analysis of trends in bird populations will not, unaided, provide the needed information. More detailed studies of particular regions or habitats or of particular species exposed to pesticides can assist greatly towards building up an over-all picture. An example of careful analysis over limited areas is the recent comparative study of bird populations in Illinois [10]. This study was based on statewide censuses, using the strip survey method, during the years 1956–58. The study was similar to surveys made in 1906–1909 by biologists of the Illinois State Laboratory of Natural History.

The study showed a loss of avifaunal variety in Illinois. The estimated total number of birds in Illinois in 1957 was about the same as in 1909—at a level of 61 million—but whereas in 1909 there were 18 common species constituting 70 per cent of the state population, in 1957 only nine species constituted 70 per cent of estimated breeding populations for all habitats accounted for in the study. This reduction in variety of the avifauna is attributed to reduced floral variety, as a result of man's activity, and to the adaptation of particular species to managed habitats. The study suggests that deforestation during the nineteenth century, as most of Illinois lands came under cultivation, probably reduced the state's breeding bird population by about 38 million.

The bird population study suggests that increased use of pesticides may be a factor in the decline of the variety of orchard avifauna involving tendencies of "insectivorous and frugivorous types such as the Eastern kingbird, robin, blue jay, and mimids and the orchard oriole" to disappear from this habitat. However, the change in the typical orchard tract from small family plots to large commercial stands offering much less edge and a more uniform aspect—far less inviting to certain species of birds—is also noted as a probable cause. Current statistical work of ecologists relating to population stability as affected by the number of food species available may be helpful in explaining these changes [11].

The Illinois study notes the sharp increase in acreage treated with insecticides and herbicides in recent years, especially in cornlands and other cultivated lands. If pesticides do cause mortality among species adapted to this type of habitat, or reduce nesting success through either loss of natural insect foods or the intoxication of juvenile birds or adults, "the very species of birds that have been increasing in recent years may in coming generations, be the species that decline." The authors confirm this possibility by references to published work, but point to the inadequacy of information on sublethal doses and long-term effects of pesticides on whole populations, especially in the light of evidence that some widely used chlorinated hydrocarbon chemicals reduce the reproductive capability of birds. They conclude that "among the most important research needs in the state are detailed, carefully planned and controlled studies of the effects of pesticides (insecticides and herbicides) on birds, their diets, their reproduction, their survival."

VALUATION OF WILDLIFE EFFECTS

From the preceding review of literature it is evident that we are not in a position to make a quantitative assessment of the aggregate effects of pesticides on wildlife even in physical terms. We may, however, with a modest program of additional research be able to determine limits within which no significant spillover effects upon wildlife occur. That is to say, it should be possible to determine, for various kinds of habitat, the dosage or cumulative annual dosage rates of each major pesticidal chemical within which no significant mortality or permanent damage to wildlife populations will be incurred.

There is some preliminary support for the hypothesis that it will be possible to find a no-effect dosage rate for many chemicals. Such indications include the fact that most cases of mortality observed among wildlife have followed either misuse of pesticides (involving application above recommended rates, or inadvertent overlapping of treatments, or drift of spray so that some areas received more than the intended amount of chemical, or use of a chemical contrary to the distributor's instructions) or special non-agricultural programs in which heavy dosage rates have been employed for the eradication of forest pests or recently imported species of pests such as fire ants. This may be, as the President's Science Advisory Committee's report [5] observes, simply a reflection of the fact that "losses following agricultural operations are more scattered and less well documented." It may also reflect to some extent a learning process in the use of pesticides. The hazards of adverse spillover effects upon fish and wildlife were largely unappreciated until they had been graphically demonstrated by fish and bird kills. It has been claimed that the major problems of wildlife losses occurred in the early years of forest treatment and that earlier mistakes have been largely corrected. [12].

It would, nevertheless, seem overcomplacent to conclude that no adverse effects result other than from deliberate misuse or gross carelessness in the use of chemicals and that the solution is simply, by education and some associated investigating and supervisory activities, to eliminate such malpractices. Strict adherence to this contention would involve defining malpractice with hindsight. Policy-making in this field will, of course, inevitably involve continuous adjustments in the light of experience. Although inescapable, it is nonetheless doubtful whether policy-making based on regrets is sufficient.

It is, furthermore, unlikely to be satisfactory to place no value on harmful consequences to wildlife on the grounds that these are small in relation to other ecological disturbances introduced by man. Simply because other sources of damage to wildlife (for example, land clearing, drainage of swamps, clean cultivation methods, multilane highway construction, or the practice of keeping domestic pets) are of possibly greater importance does not warrant neglecting the effects on wildlife altogether in the decision model for pest control.

Even if "no-effect-on-wildlife" rates of application and practices for major pesticides can be found, it does not necessarily follow that these rates should be recommended or enforced. They may, in many cases, be high enough for no conflict between agricultural and forest pest control objectives and wildlife consideration to arise. In other words, effective pest control may be achieved at rates of application that pose no threat to wildlife populations. Research to determine the maximum levels at which no effects on wildlife are observed could indicate particular chemicals or specific uses of chemicals for which no economic problem arises. Where such conflicts do arise, however, it is necessary for rational decision making to place valuations on wildlife effects.

In doing this, the problem takes a somewhat different form from that encountered in evaluating the effects of pesticides on human mortality and morbidity. In dealing with the effects on human health, a part of social benefits or costs consists of the changes in mortality or morbidity and valuation problems consist of determining an appropriate value for a life or for working time saved or lost. In the case of wildlife, however, minimum mortality of fish, birds, and other wildlife, due either to intoxication with pesticides or deprivation of food sources, is not necessarily the objective. Initial mortality may sometimes have little permanent effect on wildlife populations because of the capacity of the species for regeneration.

Furthermore, mortality caused by pesticides may merely precipitate losses of wildlife which would otherwise occur from other causes. For example, forest pests may threaten to destroy the habitat of birds and wildlife and, even though pesticides used to preserve standing trees occasion mortality among some species, the ultimate effect may be to maintain the populations concerned at a higher level than if no pesticide were applied. As an entomologist has pointed out, "the ultimate effect of insecticides on animal life cannot always be measured in terms of initial mortality of individual species" [13]. Insect depredations or insect-borne diseases of plants may well cause much greater permanent losses of wildlife as a result of the ecological disruption they initiate than occurs because of the incidental exposure of wildlife to toxic pesticides during control measures.

Moreover, possible heavy mortality of wildlife in special campaigns to eradicate or confine introduced insects or weeds before they become widespread pests may in the long run have much less damaging effects than more half-hearted initial measures. If the introduced pest is allowed to spread, the use of large amounts of pesticidal chemicals over a very much greater area may become necessary as a routine control measure. The relatively heavy application of insecticides in earlier measures to control the imported fire ant, with unfortunate side effects upon wildlife, should perhaps be viewed in this context. Use of lures and baits in conjunction with pesticidal agents has recently offered an effective means of control without the degree of spillovers previously involved.

It will, therefore, frequently be necessary in pest eradication programs to balance possibly severe short-term and local injury to wildlife populations that may be involved in drastic programs to eliminate a pest against the possibility of more widespread, less intensive, but, in aggregate, far greater damage arising from the need for control measures over a much greater area.

Even if the outcomes of alternative pest control policies in terms of their effects on wildlife were known, there would, of course, still be major difficulties in ascribing social values to the different outcomes. A number of efforts to place money values on wildlife resources have been made. One example is the National Survey of Fishing and Hunting conducted by the U.S. Department of the Interior in 1955 and 1961 [14, 15]. The values imputed to wildlife were the estimated amounts spent on wild forms by hunters and fishermen in pursuit of their hobby. This was put at about $4 billion in 1960. The values derived in this way are admitted to be incomplete, the comment being added that wildlife resources are all the more economically important when we consider the vast aesthetic values not measured in the survey.

This technique of deriving market values for recreation centered on wildlife resources uses, in effect, the value of goods and services purchased to permit enjoyment of these resources as a proxy for the value of the wildlife itself. While this procedure may provide a useful indication of productive activity supported by wildlife resources, it would be a very imperfect basis for deciding public policy questions affecting the preservation of a particular species. Any attempt to use this method in evaluating the effects of pesticides upon wildlife would have a number of anomalous implications. For example, protected species, for which hunting expenditures are illegal, would by implication be valued at less than species offering unrestricted sporting activities. This implication is, of course, incorrect.

Some extremely intractable index number problems also arise. In order to obtain an aggregate valuation for wildlife effects it would be necessary to combine outcomes for a very large number of species into a single value. If, for example, it were established that current pesticide practices had a neutral or zero outcome in regard to 400 species of birds, increased numbers of another 80 species, but adversely affected 20 other species, how would society value these outcomes? Standard weighting techniques in aggregation by index numbers are based on a linear preference scale. It is doubtful whether social values of different avifaunal species conform to such a scale. While it is possible that an extra million yellow-bellied sapsuckers would compensate for the loss of one hundred bald eagles, what value in terms of yellow-bellied sapsuckers would be placed on extinction of the bald eagle?

Values placed on birds by individuals and groups within a society tend to be spread over a wide spectrum, if not almost completely polarized. At one extreme we have the view that the "curving wing of a bird in flight" should be treated as a supreme value and that the death of a bird resulting from pesticides violates the "inalienable" rights of bird watchers to seek out subjects for observation. At the other extreme we seem to have many people willing to contemplate large-scale casualties among wildlife with complete equanimity, provided primary benefits accruing to agriculture or to human health exceed zero. Although actual social values must lie between these extremes, they clearly cannot be gauged by any kind of averaging process or opinion polls.

Value scales are perhaps more meaningfully expressed in action or decision making than as vocalized abstractions. Even the most singleminded bird lover could well revise his value scale when confronted with the full set of payoffs in a decision-making situation. For this reason, even if wildlife effects had finally to be treated as incommensurables, it would be desirable to calculate, if possible, the sensitivity of the outcome in terms of agricultural or health consequences of varying levels of mandatory protection for wildlife species. In other words, the opportunity costs of constraints imposed should be assessed, since what is originally specified as a constraint (a requirement which "must not be violated at any cost, however high, or with any probability, however low") may well be no longer regarded as such when the full extent of the effects on other objectives is appreciated.

REFERENCES

References Cited in Text

1. National Academy of Sciences—National Research Council, *Pest Control and Wildlife Relationships, Part I, Evaluation of Pesticides—Wildlife Problems,* Publication No. 920A (Washington, D.C., 1962), pp. 7, 28.
2. National Academy of Sciences—National Research Council, *Pest Control and Wildlife Relationships, Part II, Policy and Procedures for Pest Control,* Publication 920B (Washington, D.C., 1962), pp. 9, 53.
3. U.S. Department of the Interior, Fish and Wildlife Service, *Bureau of Sport Fisheries and Wildlife, Pesticide-Wildlife Review, 1959,* Circular 84, September 1960, pp. 4, 36.
4. U.S. Department of the Interior, Fish and Wildlife Service, *Pesticide Wildlife Studies—A Review of Fish and Wildlife Service Investigations During 1961 and 1962,* Circular 167, June 1963, pp. 8, 109.
5. President's Science Advisory Committee, *Report on Use of Pesticides,* May 15, 1963 (Washington, D.C.: Government Printing Office, 1963), p. 25.
6. George J. Wallace and Richard F. Bernard, "Tests Show 40 Species of Birds Poisoned by DDT," *Audubon Magazine,* Vol. 65, No. 4, July–August, 1963, pp. 198–203.
7. G. E. Burdick *et al.,* "The Accumulation of DDT in Lake Trout and the Effect on Reproduction," *Transactions of American Fisheries Society,* Vol. 93, No. 2 (April, 1964).
8. William F. Durham, "Pesticide Residues in Foods in Relation to Human Health," *Residue Reviews,* Vol. 4, 1963, pp. 33–81.
9. Frank E. Egler, "Pesticides, In Our Ecosystem," *American Scientist,* Vol. 52, No. 1, March 1964, pp. 110–36.
10. Richard R. and Jean W. Graber, "A Comparative Study of Bird Popu-

lations in Illinois, 1906–1909 and 1956–1958," *Illinois Natural History Survey*, Bulletin, Vol. 28, Article 3 (October, 1963), pp. 383–528.

11. K. E. F. Watt, "Comments on Fluctuations of Animal Populations and Measures of Community Stability," *The Canadian Entomologist*, Vol. 96, No. 11, November 1964, pp. 1434–42.

12. G. C. Decker, "Facts on Pesticide Usage," Paper to Midwest Shade Tree Conference, Chicago, February 13, 1963, p. 15. Mimeo.

13. G. C. Decker, "Insecticides as a Part of the 20th Century Environment," Ecological Society of America (Bloomington, Indiana), August 25, 1958, p. 11. Mimeo.

14. U.S. Department of the Interior, Fish and Wildlife Service, *National Survey of Hunting and Fishing, 1955*, Circular 44, 1956.

15. U.S. Department of the Interior, Fish and Wildlife Service, *National Survey of Hunting and Fishing, 1960*, Circular 120, 1961.

Other References

Governor Edmund G. Brown's Special Committee on Public Policy Regarding Agricultural Chemicals, *Report on Agricultural Chemicals and Recommendations for Public Policy* (Sacramento, California), December 30, 1960, p. 35.

James A. Crutchfield, "Valuation of Fishery Resources," *Land Economics*, Vol. 38, No. 2, May 1962, pp. 145–54.

Illinois Federation of Sportsmen's Clubs, "Let's Get Chemical Poisons Under Control," *Illinois Wildlife*, Vol. 18, No. 2, March 1963, p. 9.

Albert C. Worrell, "Pests, Pesticides and People," *American Forests*, Vol. 66, No. 7, July 1960, pp. 39–81.

8

Public Policy and Research Needs

If there is, in fact, a problem associated with the ever growing quantity of pesticides introduced into the environment annually that warrants the attention of policy makers at the state or national level, then our review has shown that there must be spill-over effects of such magnitude and distribution that the market-place cannot adjust the values. The result of this inability to make the adjustment is the improper allocation of resources. Policies are then needed to bring about the value adjustments and the proper allocation of resources in pest control.

Evidence presented indicates that the diversity of materials is increasing with time; that the quantity of material used has increased since 1939; that the trend has been away from .the arsenicals toward the chlorinated hydrocarbons, the organo-phosphates, and the carbamates; that the use of herbicides is growing rapidly; that pesticides in agriculture are possibly very productive in terms of output increases and resource savings; that effects on human health are uncertain with possible benefits from control of disease-bearing insect vectors weighed off against known hazards as well as potential hazards; and that there is uncertainty concerning the balance of gains and losses with respect to effect of pesticides (as currently used) on fish and wildlife.

We have suggested that to conclude the existence of a problem based upon such surface objectives as minimizing fish kills, birds, etc., since this is all that can be defended now, represents partialization of the more general problem that involves social welfare. To this end, the solution that will be the most helpful for policy will specify the alternatives—physical, biological, chemical, economic, and social—which relate the particular problem to the more general problem of maintaining an environment in which the goals of society tend to be achieved.

While we have not shown, through rigorous empirical analysis, that the divergence between private and social optima is of sufficient magnitude to support the conclusion that there is a pesticide problem, our review of the technical literature presents evidence that suggests the existence and nature of a problem. The very nature of synthetic organic pesticides, i.e., persistence, wide spectrum toxicity, biological concentration, etc., and the methods of applying them, provide the basis for technological spillovers and indivisibilities which may call for policies designed to reallocate resources.

An economic framework is available for analysis of the effects of pesticides on social welfare in areas where monetary values serve as resource allocators. Effects of alternative programs of regulation and public use can be evaluated in terms of implicit values generated by the activities that are governed by monetary values. This approach is extremely useful where alternative policies are designed to protect or reduce the erosion of such extramarket values as human health and fish and wildlife. Not only does such a model and approach guard against giving the wrong answers to the wrong questions, but it offers the only hope at the present time of giving empirical content and systematic analysis to the whole question of pesticide policy.

Our review has indicated a number of areas in which data improvements and additions would improve the quality of decision making. Assessment of the values and probabilities of outcomes of pest control actions, both in advance of decision and in retrospect, requires additional technical and statistical information concerning the biological effects of pesticides. Some observations on the nature of research needs for this purpose are presented as a part of this concluding chapter.

If pesticide policy is dominated or unduly influenced by special interest groups, unnecessary costs may be imposed on society. Policies and research designed to eliminate adverse effects upon fish and wildlife may represent to some the optimal solution. We have stressed the alternatives that are available of which minimizing or elimination of adverse spillovers is only one, and it may not be the most economical one in terms of national resources and values. Enlargement of the decision-making unit may be preferable so that, with much more information, decision makers can guard against hazards that are highly probable and can be cheaply prevented and can, possibly, accept other hazards as part of a more desirable outcome. Externalities can be internalized and arrangements can be made to compensate the victims of adverse effects resulting from pesticide use.

An hypothesis advanced in Chapter 4 stated that modern chemical control of weeds, insects, and fungi diseases and harvest-aid chemicals, such as desiccants and defoliants, form an integral part of the technology and economic organization of American agriculture.

If this hypothesis can be substantiated, the determination of gains from the use of pesticides in agriculture must recognize the complementarities between inputs and pesticide technology as one of a number of factors shaping the structure of the American farm. Changes in public policy that drastically alter this technology could rescind recent advances in farming methods if they reduce the effectiveness of other inputs such as mechanical cotton pickers, if they occasioned disruptions of current interregional comparative advantage, or if they significantly affected the optimal scale of farming operations. If such changes are involved in a decision, they should not be initiated by policies based on guesses.

The pesticide problem forms a part of the more general pollution problem, but is not exclusively a problem of residues. Even if all pesticides degraded rapidly and did not result in lasting contamination of the environment, external effects of some kinds would remain (for example, drift of phytotoxic chemicals, kills of beneficial insects and plants, occupational illness). Indivisibil-

ities associated with biological controls would still pose a problem
demanding government action or action by some cohesive group
to achieve the most efficient method of pest control.

Some excellent reviews of the professional literature on the
effects of pesticides on human health are available. They indicate
that storage of pesticides in the bodies of members of the general
population in the United States has not continued to increase
with the growing usage of farm chemicals. No known cases of
poisoning of consumers have occurred from residues of pesticides
applied in accordance with approved practices, although some
have followed disregard of instructions by agricultural producers.

The U.S. safety record with pesticides is not an unimpressive
one. Any obvious faults in public policy were not so much of
failure to take action to safeguard consumers against hazards of
toxic residues as of insufficient steps to make the public aware
of what the government was doing and why.

Public Policy

It would be inconsistent with our main thesis to append a list
of recommended adjustments in public policy to this exploratory
study of the needs and scope for economic analysis of the consequences of pesticides. Nevertheless, a number of questions per-
taining to policy do emerge from our study.

In the first place, the adequacy of safeguards against occupational illness among persons handling or applying chemicals is
put in question by reviews of the effects of pesticides on human
health. Cases of occupational poisoning have become more frequent. The 1,100 reported in California alone in 1959 represented
a three-fold increase in 5 years. Moreover, any signs of subtle
pathological effects of prolonged exposure to subacute doses
can be expected to show up first among such workers whose
degree of exposure is far greater than that of the general
population.

A close watch on medical histories and work logs of such
workers can provide advance warning of any unsuspected threats
to human health. Both in order to derive the benefit of this source
of information and to protect agricultural and other workers

against injury, an expanded program of regulation and medical supervision of persons engaged in the formulation and application of pesticides merits early consideration.

Second, the system of registration of pesticides and tolerances for residues appears, superficially, to be soundly based and administered. The recent steps to amend legislation to discontinue protest registrations removes one possible source of weakness. Most protest registrations have concerned new formulations of well-known basic chemicals and have usually stemmed from legitimate motives of saving the often sizeable expenditures involved in repeating tests similar to those conducted earlier for other formulations of the same active ingredient. Nevertheless, the provision did constitute a loophole in the system of safeguards.

It is impossible to evaluate the adequacy of the facilities for monitoring residue levels in foodstuffs without further careful study. However, there are grounds for concern over the limited testing of foodstuffs conducted by some states with respect to commodities produced, traded, and consumed wholly within the state.

It is true that regulation and monitoring of residue levels is only one, and perhaps not the most important, line of defense against injury to consumers. Proper labeling of containers and the education of pesticide users concerning proper methods, quantities, and time of application may well be a higher order requirement. They are, however, never likely to be 100 per cent effective, and testing foodstuffs for compliance with residue tolerances is not only a useful supplement to other measures, but also can provide information for use in policy formulation about the extent of chemical contamination of food supplies.

RESEARCH NEEDS

While it is often easy to point to shortcomings in policies and to criticize the information on which policies are based as inadequate, such a critique is unfinished unless it provides guidance to explicit research problems that will yield the kinds of information needed and suggests operational approaches to these problems. This is especially true of our study since its central theme is the objective and systematic analysis of the information that provides a basis for pest control policy.

Previous chapters have emphasized the lack of information about certain types of pesticides and have, therefore, emphasized the uncertainty which surrounds the pesticide question. It was pointed out that the uncertainty springs from the use of pesticide chemicals by millions of individual users, from the contact with these chemicals and their residues by millions or perhaps billions of organisms including man (with little or no control over this contact), and from the multitude of purposes for which the chemicals are put to use.

In such a situation it is hopeless to observe the levels of use and the beneficial and detrimental effects for each organism or set of organisms affected. It is, therefore, unlikely that any policy that is forthcoming and deals with pest control by chemicals will have the characteristics of a Pareto optimum—that is, where part of society is better off and no one is worse off. Rather, the policies are more likely to benefit some at the cost of others and it is doubtful that all losers can be adequately compensated. This does not mean to say that society cannot be made better off by more informed policies concerning pesticides. It is merely an expression of the complexities of reality that characterize the situation.

Assuming the foregoing statement to be an accurate representation of the world, research activities need to simplify the complex real world and bring some order out of chaos. This does not constitute a license to oversimplify. It is a charge to locate the most serious aspects of the problem and deal with them in such a way that incremental increases in certain values outweigh incremental losses in other values. These valuation problems will involve the co-operation of the broad scientific and political communities. These are questions not only for biologists, or chemists, or economists, or politicians, but for all of these disciplines in concert.

Research has the objective of creating knowledge, and the result, therefore, is amenable to generalization. This means that the result should provide answers to the questions directly or the research result should be an approach that can be used over and over in new situations. It should be clear at this point that research results which answer questions concerned with cotton and pesticides in Alabama cannot be applied even to cotton culture in California. Likewise, research showing the effects of pesticides on fish and wildlife in the Chesapeake Bay cannot be applied to fish and wildlife generally, nor can the effects of pesticides on formulators be applied to humans generally. The dosages differ, the chemicals differ, and the context in which use is effected is different. The problem becomes enormous.

We have spoken already about one approach. That was to develop controls free of external adverse effects and to prohibit the use of controls that exhibit potential external adverse effects. We have also shown why this approach need not be optimal.

Our approach is this. First, locate those sources of contamination by pesticides where the payoff resulting from pest control is greatest. This would also identify those sources resulting in marginal payoffs. Second, match whatever evidence is available concerning adverse effects with the sources of contamination and give highest priority for research designed to achieve control and reduce adverse effects to those areas where large benefits from pesticides seem to be related to large external costs due to adverse effects measured by whatever criteria are available. For those areas where rather small benefits from pesticide use appear to be associated with sizable external costs, assign a lower priority for research and so forth. The matrix in Table 20 makes the

Table 20. Assigning Research Priorities Based on Correlation of Benefits and External Costs

External costs	Benefits	
	Low	High
LOW	Priority 4	Priority 3
HIGH	Priority 2	Priority 1

research priorities clear. The development of such a matrix for all sources of contamination—industrial, office and home, and agriculture and forestry—then provides guidance for the research investment and can be used to develop the second round of more detailed interdisciplinary research needed to deal with the problems. These studies would delve into the effects of pesticides on the ecology and the economic impact of systems of pest control designed to reduce the adverse effects.

Research Designed To Set Priorities

We have mentioned above the sources of contamination. These we designated as industrial, home and office, and agriculture and forestry. It can be argued that the industrial exposure of humans can be controlled by a step-up in industrial hygiene efforts. In the case of home exposure, since home owners can choose the chemicals used, stricter registration laws on compounds, coupled with a vigorous education program, could markedly reduce human and wildlife exposure to chemicals that are known or thought to be harmful. As for exposure at the office, municipal health departments can develop means of reducing human exposure by prohibiting use of certain harmful or suspect compounds used to maintain building sanitation.

Agricultural pesticides have been estimated to account for about 60 per cent of the pesticide chemicals used annually. The rates of application, the chemicals used, and the economic impact are variable from region to region. Based on data obtained from the Farmer Cooperative Service of the U.S. Department of Agriculture, the average rate of use per acre of 59 principal crops harvested in 1963 varied from $.15 in Kansas to $20.02 in Florida. Further, 16 states representing most of the corn belt, most of the great plains, and including Tennessee and Nevada, accounted for 48.7 per cent of the land in farms in 1963. Yet they were responsible for only 14.7 per cent of the estimated pesticide expenditures. On the other hand, 16 states representing most of the Atlantic seaboard, the Mississippi Delta, Arizona, and California accounted for 21.7 per cent of the land in farms in 1963, but accounted for 56.5 per cent of the pesticide expenditures.

One hypothesis is that research should first focus on the pest–pesticide-production relation in these high use intensity states and in other states where a similar agriculture is found, such as in the Rio Grande valley in Texas. At the same time preliminary assessments should be made of the pesticide–human relationships and the pesticide–fish-and-wildlife relationships in these areas. If these preliminary assessments reveal that external costs are highly probable and are of significant magnitude, then the research can be directed in a more detailed manner toward the specific areas, crops, chemicals, and adverse effects that are represented. This step will be discussed later.

In a similar vein, one hypothesis is that if the cereal-producing sector of agriculture results in relatively low levels of environmental contamination from pesticides and/or the use of these chemicals does not appear to be a necessary adjunct to profitable production of this class of food, then the pests involved and their control do not receive a high research priority.

The philosophical position of this strategy is based on the economics of information gathering. Very simply, it says that while it would be ideal to know everything, some information is more valuable than other information and the cost of obtaining information should be less than or equal to its value.

One procedure for making the preliminary assessments of the importance of pesticides to agriculture in the various regions is to estimate input–output relations for regions, relating pesticide inputs and other inputs to agricultural output. These relations would provide estimates of the relative contribution of pesticides to agricultural output in the regions as well as the relative contributions of pesticides and other inputs.

Data necessary for estimating these input–output relations are currently being developed by the U.S. Department of Agriculture as the result of a survey of several thousand farms in 1965. Information obtained from farmers gives materials used, rates of application, cost of application, acres of animals treated, products treated, and farm organization. Data by states are also available currently from the USDA.

Another procedure involves the development of regional

budgets providing a systematic accounting of output and resources used. With this kind of synthesis and the help of pest control experts in the development of reasonable assumptions concerning yield increases and frequency of damaging infestations,
the sensitivity of output to incremental changes in pesticide applications can be examined. In addition, the effect of such output
changes on consumer outlays and resources required could be
estimated along the lines presented in Chapter 3. Data required
include a prescription of pesticide application for each major
product, output levels, and inputs of land, labor, and capital.

While the approaches just mentioned are subject to limitations
and rather gross aggregation, they appear to be, in the absence
of a sophisticated damage function for the various pests, the most
promising avenues of approach at the present time. A great deal
of information concerning data needs and priorities for more
detailed research in the chemical and biological areas should be
provided by such an analysis.

Detailed Research on Problem Areas

In the previous discussion we presented a method for assigning research priorities to problem areas. When these have been
identified, studies can be designed to attack the pest control
problems.

Because of the numerous interactions of pesticides with the
elements of the environment, this second-stage research will need
to be very detailed and cover the total system that is affected.
Data will be needed concerning the relation between pesticide
application and pest control and this information will need to be
translated into the effects on agricultural production of specific
crops. Data will be needed on the fate of the chemicals in the
environment as well as their effect on human life, wildlife, and
fish. Questions concerning the effects on population dynamics of
species in the ecology resulting from pesticide use will require
answers. All of this information is necessary in order to evaluate
the impact of alternative pest control systems on the values both
monetary and nonmonetary that are present there.

Research of the type just referred to requires the co-operation of biologists (including ecologists), chemists, entomologists, and economists, and may involve political scientists. The objective of these studies would be the management of the affected ecological system in such a way that the values important to men are brought into some sort of acceptable balance. The authors are aware of only one such study that has been proposed that is designed to provide this kind of information.

The cost of these ecological system studies is high. Competent ecologists trained in the techniques of biomathematics are not numerous. Competent analytical chemists are also scarce. Both of these skills are vital to the type of research proposed. This is a further reason for a systematic means of assigning priorities to allocate research resources.

Research of the type suggested can answer the questions that seem to us to be important. Control of pests involves complex relationships that may not be meaningfully generated by laboratory experiments. For instance, it is not enough to know that DDT or other organochlorine products lower fertility in a certain species of fish. It is also necessary to know whether the concentrations of the chemicals in water resources that result from usual rates and methods of use are likely to reach levels endangering fish reproduction and if so to what degree and, further, what this means concerning the stability of the aquatic ecology. Long-term, as well as instantaneous, effects also need to be spelled out, accompanied by their implications for direct costs of pest control over time and associated external costs.

Other Research and Data Needs

It is impossible to pinpoint any area of absolute neglect in the great amount of research that has been and is being carried out. Nevertheless, attention to nonchemical means of control and to chemosterilants and attractants seems to have received, until comparatively recently, a relatively minor part of research resources devoted to pest control. There is still a tendency to assess the potential results of such research as limited because of the limited number of important and fully effective successes in the past. As noted earlier, this may be misleading as a guide to future returns from an intensification of research resources in this area. There may be important economies of scale in this area and the indivisibilities and frequently nonmarketable nature of research findings should not be allowed to lead to suboptimal inputs of research resources.

Since research dealing with the reduction or elimination of adverse effects from pesticide use could have major implications for the pesticide-producing industry, research studies on the structure and the competitive nature of the agricultural chemical industry are needed. This information would provide a basis for assessing the impact of changes in pest control policy on the chemical industry.

Certain steps can be taken immediately at apparently low cost to improve data which are now being collected. In order to monitor the effects on human health, a more detailed breakdown of mortality and morbidity statistics for poisoning should be made. Figures for deaths from accidental and occupational toxication from pesticides are currently grouped under a general category of accidental death from poisonous substances, solid and liquid. In view of the growing public concern with pesticides, a separate category for them in mortality statistics would seem to be warranted. In addition, attempts to compile comprehensive records of nonfatal illnesses due to accidental, intentional, and occupational exposure to pesticides would appear to merit consideration. In both cases, breakdowns by specific chemicals or groups of chemicals, as far as possible, would be useful.

The major advantage of employing a formal economic model in the analysis of the costs and benefits to society of alternative actions affecting pest control is that it compels, or at least encourages, consideration in the decision-making process of the full range of alternatives open, including some which may not otherwise be recognized as alternatives. There is, as we have seen, a very great number and variety of possible actions to deal with the problem. In all probability no single course of action will be sufficient in itself to yield the best possible outcome for society.

There are plenty of panaceas being hawked around the policy arena; nothing will serve better to expose their limitations and to determine the real extent of their usefulness than economic analysis of the kind here proposed.